The Man Who Gave Back

Of great riches there is no use,
except it be in the distribution;
the rest is but conceit.

Francis Bacon

The Man Who Gave Back

The Rule of Life of H.R. MacMillan

Robert D. MacRae

6000 Iona Drive
Vancouver, BC V6T 1L4

ISBN 10: 1-895018-35-8
ISBN 13: 978-1-895018-35-6

Library and Archives Canada Cataloguing in Publication

MacRae, Robert D., 1932–
The man who gave back : the rule of life of H.R. MacMillan / Robert D. MacRae.

ISBN 978-1-895018-35-6

1. MacMillan, H. R. (Harvey Reginald), 1885-1976. 2. Philanthropists—Canada—Biography. 3. Businessmen—Canada—Biography. 4. Industrialists—Canada—Biography. 5. Foresters—British Columbia—Biography. I. Vancouver School of Theology II. Title.

HV28.M325M34 2007 361.7'4092 C2007-905549-4

Design and Page Composition: Jay Tee Graphics Ltd.
Cover Photography:

Printed and bound in Canada
9 8 7 6 5 4 3 2 1

Contents

Dedicated to the memory of

HARVEY REGINALD MACMILLAN
1885-1976

Philanthropist

Acknowledgements

At the conclusion of twenty-one years as rector of the Church of Saint John the Divine in Victoria, British Columbia, I was granted a sabbatical leave during which this account of the philanthropy of H.R. MacMillan had its beginning. While there is some public record of Mr. MacMillan's considerable gifts to the community, I was curious as to what motivated him to become British Columbia's first major philanthropist.

There are many people to whom I am grateful for guiding me in this adventure: John Oussoren of the Chalmers Institute at the Vancouver School of Theology (VST), who encouraged me in self-directed studies in planned giving. Several archivists spared no effort to help me: Ann Watson, VST; Doreen Stephens, the Anglican Provincial Synod; Bob Stewart, the United Church BC Conference; and George Brandak, UBC Manuscripts Curator. Herbert O'Driscoll, former Dean of Christ Church Cathedral in Vancouver, gave me the encouragement I needed when it seemed this work might never be completed; Jim Taylor shared reflections of his father W.S. Taylor plus wise editorial suggestions; William Gibson shared remembrances of his friend H.R. MacMillan and of their dream that was never realized.

From a great distance came more help: Joan Dutson, Archivist at the Laurentian Institute in Montreal and Joan Shaw of St. Andrew's Presbyterian Church in Aurora, Ontario provided information on MacMillan's early life.

Ken Drushka, the author of H.R. MacMillan's major biography, read early drafts of my manuscript and shared aspects of MacMillan's life I would not have otherwise known. I am grateful to Ken's wife, Laura Wilimovsky, who gave permission to quote from her late husband's book, HR. Several others provided helpful suggestions: in Vancouver,

Martin Elfert, Martha Lou Henley, Julia MacRae, Corinne Rogers, and Nancy Southam; in Victoria, Anita Hadley and Daphne Rogers. I am grateful to Megan Brand for her careful reading of the text and her insightful editorial skills. VST Principals Bud Phillips, Ken MacQueen, and Wendy Fletcher encouraged me to complete the work and agreed that the story of MacMillan's philanthropy should be shared with British Columbians. I am deeply grateful to Jean MacMillan Southam who made this publication possible.

My deepest appreciation is for Susan, my wife, best friend, and *anam ċara*, from whom I daily receive unwarranted love and patient wisdom.

Robert MacRae
Victoria, BC
August 2007

The End is the Beginning ...

What we call a beginning is often the end
and to make an end is to make a beginning.
The end is where we start from.

— T.S. Eliot
Little Gidding

On February 9, 1976, the giant of British Columbia's forest industry was felled. Dead at ninety years was Harvey Reginald MacMillan, for decades a leading thinker and business personality in this province. His death made headlines across Canada and around the world. Predictably, most news stories focused on his influence in making the BC forest industry the biggest employer in the province – almost everyone knew about his business acumen and achievements.

The eulogies following MacMillan's death focused on varied aspects of his life. What some considered to be his most significant contributions were reflected in editorial tributes. In the business section of the *Globe and Mail,* the editor titled his tribute to H.R. MacMillan, "The radical H.R." The writer described him as, "an entrepreneur ... a capitalist ... a genius with a dollar ... a genius with a piece of timber ... a giant in British Columbia ... a household word in the province," and went on to give him the following credits:

> It was H.R. MacMillan who brought scientific logging to the province, who wrote the report on sustained yield timber management for the B.C. government that became the foundation for the province's forest management policy. It was with this kind of expertise that he put together one of the world's largest lumber and pulp-paper complexes.

> It was this kind of mind that could lead him to urge 20 years ago that Canada recognize the People's Republic of China. Refusal, said this capitalist, was "only emotional", [it was MacMillan] who warned 20 years ago that American investment in Canada was growing faster than the Canadian economy, and blithely announced 15 years ago on national television that capitalism was on its way out.[1]

A more perceptive view came from *The Vancouver Sun* the following day:

> The fall of a forest giant makes the earth tremble far around. Men pause in awe and respect. They know the rare combination, which perhaps once in a forest's lifetime, produces the specimen that towers above all others ... He was a great patriot who knew Canada intimately and loved it. He did infinite service for his province in building BC's biggest industrial unit in which imaginative daring provided thousands of jobs. In the end he gave his money away. His interests were wide and so were his philanthropies. One of Canada's first graduate students himself, his millions enabled the University of BC to establish itself in advanced studies that will produce our future leaders.[2]

On the same day as the above editorial appeared in the *Sun*, another perspective offered a radically different approach. On February 10th, the Feast Day of Saint Scholastica, the Poor Clare Sisters of Mission, BC gathered in their little chapel in the early morning to honour a close friend of the celebrant's parents. The Reverend Robert Damasus Payne, a priest of the Benedictine Order located at Westminster Abbey in Mission, BC, stood before the altar ready to offer a requiem mass. In the course of his homily, the preacher quoted his Abbot who had declared that with the death of H.R. MacMillan, "An era is over." He chose his sermon text from the *Revised Standard Version of the* Bible (1 Kings 8:27): "Behold, heaven and the highest heaven cannot contain thee; how much less this house which I have built!" He likened MacMillan to King Solomon:

> He was a builder on a large scale. He stood taller than the trees he dealt with ... Some people won't picture him in such a glowing light. They called him a tiger. A hunter, a "true hunter" would be a better term. True hunters respect the game they stalk, because they are not alone in the woods.

> They see the signs of God's presence everywhere and know that He is running with them, encouraging them.[3]

MacMillan's biographer, Ken Drushka, described MacMillan's enormous personal impact on people as follows:

> People reacted to him far more strongly than he did to them. He was like a loaded freight train, thundering through the quiet villages of their lives.[4]

MacMillan's life and career were described by Roland Wild in an extensive article, "Tycoon of the Tall Timber," published in the March 15, 1948, edition of *Maclean's* magazine. By the time of the interview, MacMillan had come to know several Canadian artists personally and had purchased a good deal of their work. Wild described him as the "biggest man in town" (that is, Vancouver, BC) and went on to refer to him as:

> A symbolist-painter might portray him accompanied by a forest (half growing and half falling), a ship in the yards, a salmon on the line or about to enter a can, a globe of the world and a book of history. He is a complex character, unusually charming when he wants to be, unusually forthright when he decides not to be charming.[5]

A citizen of two worlds

These perceptions of MacMillan suggest to me that he was a citizen of two worlds. Was this impression, perhaps, the "rare combination" of which *The Vancouver Sun* spoke? Clearly, he moved with ease between two worlds: the sacred and secular – the spiritual and temporal.

What was it that shaped this leader of industry to become so strong a personality, such a mighty presence among people? As I hope to reveal in the pages to follow, there were many facets to his life: he was a generous, forgiving, sensitive, perceptive, and discerning man. Furthermore, as with St. Paul in his letter to the Corinthians, MacMillan "did not suffer fools gladly."

In their attempt to describe MacMillan, editors, preachers, and biographers use a wide range of words and metaphors: a radical, a hunter, a tiger, thunder, a loaded freight train, a tycoon, a ship in the yards, a book of history, and one compares him to the Old Testament personality, King Solomon. I understand biographer Ken Drushka's analysis of

MacMillan as a "contradiction" – much like Walt Whitman's description of himself: "Do I contradict myself? Very well then I contradict myself, (I am large, I contain multitudes.)"[6] On the other hand, MacMillan's intellect, his wide-ranging business involvement, and a close brush with death enabled him to shape a finely honed and simple rule of life. I believe this rule guided relationships with his family, his business and, finally, with the wider community.

Still, the life of H.R. MacMillan raises many questions: What influences molded this strong character: family, education, religion, or business? What values do we find in his character that may be said to sum up his life? Finally, what was it that prompted the generous benefactions he made toward the end of his life?

Chapter One

A Time to Be Born

Harvey Reginald MacMillan was born at home on Lot 32, Concession 41 of Whitchurch Township in York County, Ontario on September 9, 1885, and his birth was registered on November 15th. His birth at the end of the 19th century came precisely at the dawn of a new era in Canada. The political and economic leaders of the time were devoting their attention and energies to the undeveloped West. On November 9, 1885, exactly two months after MacMillan's birth, construction workers drove the last spike of the new railroad into place. Now packaged to the rest of Canada via two ribbons of steel, the West welcomed thousands of new settlers from Great Britain, Europe, and the eastern provinces. They became the railroad's first wave of passengers as they moved westerly toward the Pacific. It was an age when entrepreneurs sought their fortunes through development of the vast natural resources that lay largely dormant in the western and northern territories and especially in the new province of British Columbia that had entered Canada's confederation in 1871.

Harvey Reginald MacMillan, known as "H.R." to many people, was a second-generation Canadian by birth. In time, he came to embody the hard-work ethic and faith of his Scottish forebears combined with the aggressive entrepreneurial spirit of his period. By the mid-20th century, he had become widely known as the most successful businessman in the history of his adopted province. Today, British Columbians remember him as the pivotal person responsible for shaping the first hundred years of the forest industry in B.C. More importantly, they remember him as this province's most generous philanthropist of the 20th century.

Three major influences shaped the life and destiny of MacMillan: his family, his religion, and the opportunities provided by his country.

Harvey Reginald McMillan at age six months. Reggie, as he was known to his family, was born in his parents' farm home in Whitchurch County, Ontario, 1885.

The McMillan's two-story home was located on Lot 32 , Concession 41 of Whitchurch Township, York County, Ontario.

These powerful forces, always at play in his personality and thinking, were the ligaments that held him together. They shaped and sustained him as one of Canada's most dynamic and influential citizens of the 20th century.

The family influence

Harvey Reginald MacMillan's paternal grandfather was John McMillan and his grandmother Edith was a Willson. Grandfather John McMillan was born in Wigtown, Galloway, Scotland in 1835, the son of a Scottish Highland farmer who had fought in the battle of Waterloo in 1815. John emigrated to Canada in 1849. In the 18th and 19th centuries, immigration officers and the clerks who kept ships' passenger lists, as well as clergy and teachers, frequently abbreviated the passengers' surnames from "Mac" to "Mc." Early in the 20th century and while at college, young MacMillan became a stickler for detail and changed his surname to **Mac**Millan.

Genealogy of Harvey Reginald McMillan and Edna Mulloy

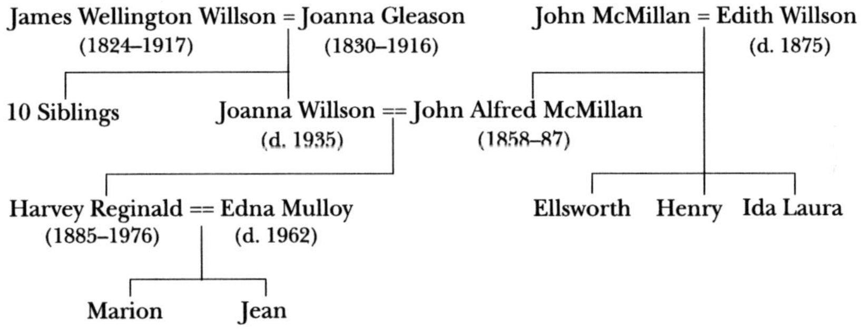

Grandfather John McMillan and Grandmother Edith Willson worked hard, saved, borrowed, and invested in a farm located on Lot 32, Concession 4, to the southeast of Newmarket, near Pine Orchard, Ontario. Three of the McMillan brothers also settled in this area. Edith died of tuberculosis after the safe delivery of four children, including John Alfred McMillan (Harvey Reginald MacMillan's father). TB, a highly contagious disease typically spread through close contact among families living in cramped quarters, ravaged many pioneer families and

Joanna Caroline Willson, H.R. MacMillan's mother, is one of six daughters of his maternal grandparents, James Wellington Willson and Joanna Gleason. Joanna is standing at the centre of the photograph taken in the early 1880s. This family also included five sons.

The three sons of John McMillan and Edith Willson: Henry (teacher), John Alfred (farmer) and Ellsworth (dentist).

Edith was blamed for bringing TB into her family. Her three sons died after contracting TB. When her widowed husband John married again, it was tuberculosis that killed at least six of his eight children from this marriage.

John Alfred McMillan and his wife, Joanna Willson, continued to operate the family farm for a short period following grandfather John McMillan's death. They were involved in the social life of the rural community and John Alfred McMillan and his extended family were active in the Pine Orchard Meeting House of the Society of Friends, the Christian community that nourished their deep personal faith. This small church was a branch of the Hicksite subdivision of the society and was located approximately half-mile east of the McMillan family home.

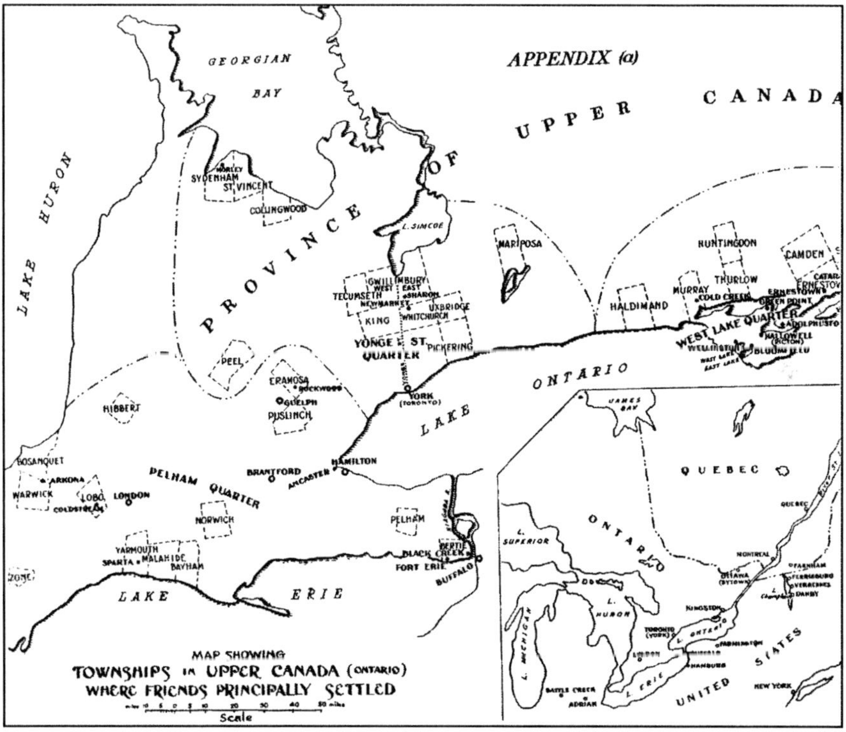

The early 19th century missions of the Society of Friends were extensive throughout Ontario and particularly concentrated in the Newmarket area north of Toronto.
Credit: *The Quakers in Canada*, A History: Arthur G. Dorland, inside front cover, The Ryerson Press, Toronto 1968.

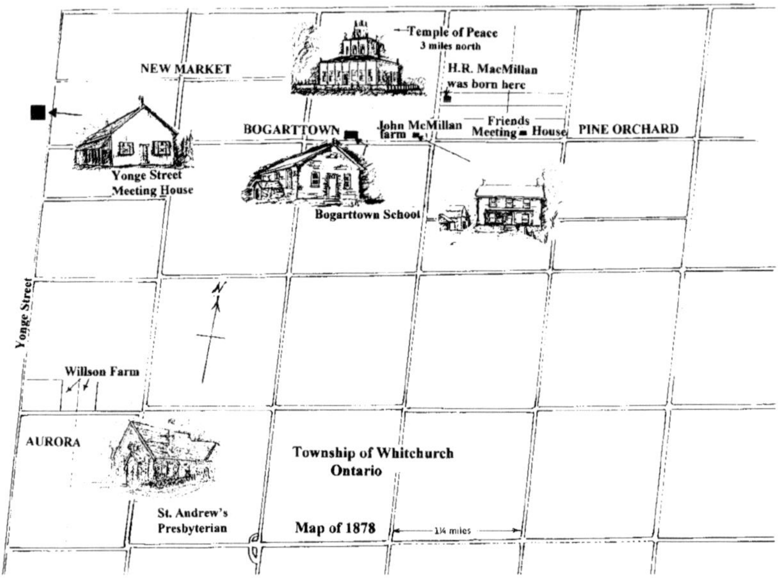

The Township of Whitchurch in 1878 showing H.R. MacMillan's birthplace and early Meeting Houses of the Society of Friends including Sharon Temple of Peace built in 1828. From time to time, the McMillan family worshipped in the Yonge Street Meeting House built in 1810 and typical of many early Meeting Houses. They also attended the Pine Orchard Meeting House. Harvey Reginald attended schools at Bogarttown and Newmarket, Ontario.

H.R. MacMillan's parents, John Alfred and Joanna, were married in 1883 in a Presbyterian Church in Simcoe County, near where her parents lived. Because there was no Quaker Meeting House there, they joined the local Presbyterian Church. The Society of Friends took a dim view of marriage outside their close-knit community. Despite the long and faithful membership of their respective families in the Religious Society of Friends, the Quakers disowned John Alfred and his bride. The Quaker discipline on marriage held that it was not dependent upon the mediation of a third person. "Not only was a person disowned for having his marriage performed by a priest, but even to be present at a ceremony of this kind was regarded as countenancing a 'hireling ministry' and was likewise a disownable offence."[1] Neither inter-denominational cooperation nor mutual recognition was acceptable to the Quaker community of that day. Despite this rejection by the Quakers, the MacMillans did not seem to hold any long-lasting negative feelings towards their beloved church and they continued to attend both Presbyterian Churches and Quaker meetings. One child was born to this marriage, Harvey Reginald.

Divisions in the Quaker House

Joanna Willson's parents had emigrated from the United States to Canada in the early 19th century seeking freedom and opportunity. They were to become a primary influence in the life of young MacMillan. Together with her extended family, they were pioneers in Simcoe County, some seventy-five miles north of Toronto. Some of the Willson family settled in York County. Joanna's grandfather, Hiram Willson, helped to build the famous Sharon Temple in Whitchurch County.

H.R.MacMillan's parents, John Alfred McMillan and Joanna Willson in 1889.

Hiram's father, Hugh L. Willson, was a stepbrother to David Willson, the founder of the Children of Peace, also known as the "Davidites," a liberal breakaway sect of the Quaker movement.

David Willson came to Canada about 1801.[2] He claimed to be raised by "pious Presbyterian parents whose fortunes in life left me far below the means of common school learning."[3] He joined the Society of Friends shortly after his arrival in Canada and was an earnest member of the society for about ten years. However, disagreements emerged on the question of war and a free gospel ministry and a breakaway group of the Society of Friends developed. David Willson, along with many others, was disowned by the Quakers and his response was to establish the Children of Peace. This group employed building techniques similar to those used by Solomon in the construction of the Temple at Jerusalem; they used no nails when they erected the unique Temple of Peace at Sharon, Ontario in 1811-12.

The Sharon Temple housed the Children of Peace's comparatively elaborate rituals and the ceremonies used in their worship rites. David Willson's break with Quaker tradition and discipline reveals something of the depth of his reaction to the simplicity and quietism he embraced when he had initially converted to Quakerism. This rift ignited when David Willson not only broke with historic Quaker views on politics, war, and ceremonies, but also when he insisted that music be a part of the worship in the Quaker Meeting House.[4] He was a prolific writer of small meditation books on subjects of a spiritual nature, one of which was among MacMillan's prized possessions.

David Willson's interest in music included the writing of many hymns. From his hand, we have the beautiful lyrics set to the tune *New Britain,* popularly used in the singing of *Amazing Grace*:

It's love the peaceful olive bears,
And love's the fruitful vine,
And love's glad tidings in our ears,
To feed these flocks of thine.

Love is the law thou didst command,
Thy children did declare;
Love is a blessing in the land,
And joyful fruit doth bear.

Love is a diadem most bright
In ancient ages shone;
Love is a star of morning light,
A saviour of mine own.[5]

The ideals expressed in the hymn grew out of the piety of the Children of Peace and the theology of the Religious Society of Friends. Some people who were outside the Quaker Church, and therefore not familiar with the deeply felt reasons for the branches within the Society of Friends, found cause to ridicule the Quakers and their divisions. One historian, Arthur Dorland, suggests that the Children of Peace "must have stimulated the imaginative and musical life of the community, and added to the gaiety of the nation, even if failing to make a very deep impression on its religious life."[6] Dorland mused, "While David Willson may have added nothing of permanent value to the religious life of Upper Canada, he undoubtedly did something in his original way to relieve the drabness and monotony of life during those early pioneer days, when the whisky bottle was the easiest and most popular means of escape from the dull round of everyday life, and when there were few legitimate means of recreation and of self-expression."[7]

Left fatherless

Unlike David Willson's limited appreciation of the value of an education, H.R. MacMillan's paternal family was committed to the pursuit of professional education and training for their children. One of MacMillan's uncles, Ellsworth McMillan, became a dentist and another, Henry McMillan, a teacher. The latter died from tuberculosis at a young age. H.R.'s father hoped to become a medical doctor and worked hard to finance his own education. Sadly, he also succumbed to TB in 1887 when "Reggie" (as he was then known), was only two years old. Young MacMillan's widowed mother Joanna, with no inheritance from her husband's estate except for ten dollars, was left to cope[8] with the help of the extended and close-knit McMillan and Willson families. As a young child MacMillan experienced what he later described as "extreme poverty." He spoke of it as follows:

> I know something about it because my father died when I was two years old, and I was brought up by my grandparents and my mother who lived in what today would be called "extreme poverty." The chief source of reve-

Harvey Reginald MacMillan about three and a half years old, 1889.

The young MacMillan at age six years in 1891.

> nue was milk, which was produced by hand and shipped about 30 miles to Toronto, for which the producer got 8 cents a quart ... I can remember my mother and her sisters worked very hard milking cows ... [and] my mother going out to milk the cows early in the morning while I was still in bed.[9]

For about three years, young MacMillan moved with his mother to different communities, sometimes living in the homes of Joanna's employers. Finally, when she could no longer keep him at her place of employment, Joanna sent him to live with his elderly maternal grandparents. Some of his uncles and aunts were also living on the farm.

The influence of MacMillan's extended family on his life, and particularly that of his maternal grandparents, was permanent. Their example of hard work and fortitude shaped the life of young MacMillan. He shared responsibility for the family chores, particularly gardening, the feeding of animals and caring for the poultry. Despite this, he hated farm work, particularly tending to the chickens, and throughout his life he never ordered a chicken meal. Nevertheless, the persistent farm chores, demanding and trying as they were, helped to shape the man he was to become. Certainly, the experience served to identify a lifestyle he did not wish to pursue.

Grandparents' support

For most of his childhood and youth, MacMillan lived with Joanna's parents, the Willsons, who attended a nearby Quaker Church. In a letter to the Religious Society of Friends in 1970, MacMillan said he was "trained to sit quietly on a bench up against the outside wall of the Meeting House."[10] Many Quakers believed that with silence and reflection, "a person is capable of extraordinary activity."[11] Perhaps this experience, what Quakers called passivity and emptiness, was the beginning of MacMillan's comfort with, and enjoyment of, reading and reflection.

MacMillan continued his schooling while living with his grandparents. He attended the nearby rural elementary school, a four-mile hike each way. Later, he attended the local high school, three miles away in distant Aurora. His grandparents moved to Aurora for his final year of his secondary schooling and this meant he did not spend as much time walking to and from high school. Although a good student, he did not spend all of his time on schoolwork. Early in his high school career, he found sufficient opportunity to court a classmate. One of his favorite

The red brick school at BogartTown where MacMillan attended elementary school. Built in 1859, it is the oldest school in Ontario outside of Toronto.

School boy Reggie MacMillan at age ten years in 1895 was always dressed in white shirt, tie, and suit.

activities while walking to school was thinking about Edna Mulloy, the daughter of the Aurora High School Principal, C.W. Mulloy. Everyday, as he trekked along, he picked a rose bud for her and then placed it on her desk at school.

The outside world

Although MacMillan's world was rural and very provincial, there were elements about it that subsequently led him to appreciate the wider world. "My neighbourhood," MacMillan once noted, "extended for a radius of how far one could drive on a gravel road with a horse and buggy in a day, and get home in time to do the chores."[12] Nevertheless, an important contact with the outside world was forged through his Grandfather Willson's regular subscription to *The Weekly Globe.* The *Globe* brought a sense of the wider world into his life, and he read it carefully. Its focus on national and international affairs helped to shape MacMillan's worldview and introduced him to the unfolding history of his times.

From time to time, young MacMillan visited members of his extended family in Toronto. His first cousin, the future writer Mazo de la Roche, (famous for her sixteen volume series on *Whiteoaks of Jalna*) was his senior by six years. She loaned him the first novel he ever read and thus began in him the joy of reading fiction and historical novels. For the next five decades, the two cousins maintained a close relationship and a regular correspondence until Mazo's death in July 1961.

Although living with his grandparents meant that MacMillan was separated from his mother, he would visit her regularly on Sundays. They would each walk to a rendezvous point midway between his grandparents' home and the place where Joanna was working, a distance of about four miles. This arrangement allowed mother and son to have time together without distractions from other relatives. Joanna was determined to keep a close and lively relationship with her son even though she was not able to live with him on a day-to-day basis. He lived with her in Toronto for a brief period and they spent holidays together and occasionally travelled to Niagara Falls or to Toronto tourist attractions. During his childhood and youth, she took every opportunity to expand his experience and understanding of the world. This positive and mutually supportive relationship of mother and son continued for decades. Indeed, when MacMillan and his wife moved to Victoria in 1912, Joanna came with them, living with them until her death on November 30, 1935.

Core values

The family held tenaciously to their core values of hard work, education, and personal morality. In addition, Quaker and Presbyterian values, such as the need to equip young people with training, self-discipline, and education reinforced their Quaker influences and were imprinted on the whole family. However, like his uncles before him, MacMillan did not wish to be a farmer. He did not want to duplicate the isolation, the hand-to-mouth existence and the marginal financial rewards that beset agricultural families of the period. Education offered a way out, and his mother and extended family encouraged him to apply to university.

What and where should he study? How could he finance an education? Here his extended family, the determination of his mother Joanna, and a chance-opportunity converged. The answer to their questions came through an unexpected meeting with Leonard S. Klinck, a representative of the extension department of the Ontario Agriculture College (OAC) in Guelph. In the summer of 1902, while young MacMillan was hoeing potatoes in his grandparents' garden, Klinck

Leonard S. Klinck, a slightly older contemporary of MacMillan at Ontario Agriculture School, encouraged MacMillan to attend OAC. Klinck became UBC President in 1919.

Photo Credit: Archival and Special Collections. University of Guelph.

arrived in a one-horse buggy and initiated a conversation with young MacMillan that moved quickly to the possibility that he might apply to the college. This important interview between Klink and MacMillan took place across the fence by the roadside.[13] Meanwhile, the McMillan family had another connection with OAC in the person of Charles Zavitz, a distant cousin (by marriage) who lectured at the college. Perhaps it was Charles Zavitz who suggested the chance meeting by the field of potatoes.

MacMillan wasted no time following these leads: he earned a little money in the early summer (his first sales job was selling stereopticon slides, a popular entertainment device) and he took a second job (offered through cousin Zavitz) for which he earned nine cents an hour tending experimental seed crops at the college. This wage is equivalent to approximately $2.07 per hour in 2007, a very low wage by any standard.

A plan for further financial support unfolded. Uncle Ellsworth McMillan offered some modest financial help to cover the sixteen-dollar annual tuition fee, (low by today's standards, but high at the time), and Joanna would contribute all she could to cover the costs of board and room.

MacMillan applied to the college and was accepted. By summer's end, and just turning seventeen years old, MacMillan was on his way to higher education. Something else put a song in his heart: just before his seventeenth birthday he had become engaged to his high-school sweetheart, Edna Mulloy. The three great influences in his life were converging: family, faith, and the Canadian opportunity.

Chapter Two

A Time to Prepare

It may seem curious that a young man who had decided he did not want to be a farmer should apply to study at an agriculture college. MacMillan's application and speedy acceptance by the Ontario Agriculture College were typical for the man who, throughout his life, would repeatedly walk through doors of opportunity. He knew education was a key to a fulfilling life, and at the time, the OAC seemed to be the only place of higher learning that was accessible to him. It would seem the hand of providence was with him and he seized the moment with enthusiasm.

The Agriculture Judging Pavilion at Ontario Agriculture School, Guelph built in 1902, the same year as MacMillan began studies there.

MacMillan applied himself to his studies. A strong and self-assured character, he was popular with his peers. A fellow classmate described MacMillan as "a very tall, rather angular and raw-boned Scot with a quiet voice, a hearty laugh and a genial personality."[1] In a word, he was a "presence" on the campus. He established a Bible study group attended by fellow students in his college residence and he was confirmed at the Knox Presbyterian Church in Guelph, during the winter of 1903. Perhaps the classes leading to confirmation helped to shape his lifelong concern with how people, including himself, should behave. However, interest in Christian morality did not deter his activities with fellow students. He was a member of the OAC hockey and football teams who had a reputation as a prankster on the campus and was often found instigating activities that were not popular with authorities: trashing students' rooms, upsetting beds, and diverting rain gutters into colleagues' rooms.[2] Although a well-rounded student, his strengths seemed to lay in the arts and humanities, particularly in literary endeavours.

One story from his student days at the Ontario Agriculture College reveals MacMillan's academic acumen at a time when he was the editor of *The OAC Review,* the student literary journal. For someone as gifted in the literary arts as he was, MacMillan's position as student editor became all-consuming, and his studies took a backseat to this job. He approached his role as editor with enthusiasm and often skipped classes to prepare the *Review* but was chastised for it by his professors.

Even though he had an uncanny ability to discern the key points of a lecture, MacMillan was not attending classes to hear his professors lectures – he missed every lecture given in a course on cold storage. "The night before the final exam he gathered a bunch of the boys in his room for a bull session."[3] MacMillan asked his colleagues to summarize the main points of the course and the following day he was the only one to receive one hundred percent in the exam.

MacMillan's facility with his pen and imagination gained him a reputation as a brilliant student of English. In his first year, he won an English scholarship that awarded him twenty dollars toward his tuition in second year. Not surprisingly, he became editor of *The OAC Review* and some believe it has never again achieved the high standards MacMillan set there while serving as editor. Still others thought his career would be as a writer or a poet for he was "an artist in words."[4] But if for a moment he thought he might follow a career in journalism or writing, that was about to change in dramatic ways.

Career choice

In 1903, as MacMillan began his university career, the administration of the OAC was beginning to develop an interest in forestry. Dr. George Creelman, the new president, encouraged this direction, while visiting speakers and campus programs began to focus on forestry which, in turn, triggered MacMillan's interest. By February 1904, he was drawn to the possibility of a career in forestry, and he sought a summer job with the Federal Forest Service. Despite the refusal of his application, the newly emerging profession continued to beckon him. Through a curious connection of extended family and college associations, a further employment opportunity changed his life forever.

Edmund J. Zavitz, a nephew of Charles Zavitz and a cousin to MacMillan, offered him a position at the OAC to establish a tree nursery on campus. Earlier, he had turned down the opportunity to undertake a similar venture in Saskatchewan but, upon arrival in Guelph, Zavitz met MacMillan and persuaded him to apply for work at the developing prairie nursery. His application was accepted and, in the summer of 1904, he was working in Indian Head in southern Saskatchewan, which was then still a part of the Northwest Territories.

Harvey Reginald MacMillan was editor of *OAC Review*, the student literary journal at Ontario Agriculture School, Guelph.

The heartland of the prairies seemed an unlikely place to find an aspiring forester at work. Across Canada, a burgeoning interest in forestry prompted the federal government to support this emerging profession. Indeed, the forest industry itself began to recognize the need to exercise stewardship of timber as a natural resource, and to systematically restore forests already ravaged by heavy harvesting. This recognition could not have been timelier. The timber industry was burgeoning with the rampant growth of the nation, and the need for housing and the demand for lumber for all kinds of construction led to a building boom. Furthermore, both the paper industry and the expanding export trade in lumber demanded more resources of the forest – for the raw resource and the finished product. MacMillan's summer employment in western Canada introduced him to the vast expanse of the country, its forest assets, and to the future potential of this untapped resource for an expanding population.

In the summer of 1905, MacMillan took a position in Manitoba working for the Forestry Branch of the federal government in the Turtle and Moose Mountains near the Manitoba border with the United States. Here, he led the first survey of forest reserves ever conducted by the branch, and he thrived on a full summer of outdoor living. He loved everything it entailed: camping, hunting, and fishing. This period marked the beginning of his lifelong affection for the natural environment and his summer experiences convinced MacMillan to pursue a career in forestry.

Some sixty years later, MacMillan penned a letter to Zavitz describing his first summer in the prairies and what it meant to him:

> When you dropped the idea (of going to Indian Head), Creelman mentioned to me that it might be a good thing for me to do. I would get my fare paid out and back; I would be given $25 a month and board (I think) for the summer, in addition to which it would give me a chance to see something of the West and get acquainted with the Forest Branch of the Department of the Interior and perhaps get a start with them. This turned out to be the turning point in my life.[5]

Moral and ethical awareness

In his biography, *H.R.*, author Ken Drushka suggests that MacMillan's religious interests "probably focused more on the moral teaching of the church than on spiritual matters."[6] Early in his career, he began to con-

"Living the simple life" is the way MacMillan described his summer experience in Riding Mountain, Manitoba in 1906. He is seen on the right in the foreground.

sider the moral implications implicit in the exploitation of forests, and this concern continued in his later professional and business life. His developing sense of the need for a committed stewardship of the forests was evident when he returned to the OAC in the fall of 1905. Now in his final undergraduate year, he wrote a thesis, *Forest Conditions in Turtle Mountains,* based on his summer's research and observations. He was critical of the damage caused by careless fires, of logging practices that were wasteful, and of the intervention of governments. He thought that forest squatters, camped in wilderness forests, held contempt for order which, in turn, constituted a "source of moral danger in the community."[7] If forests were not protected, he felt, serious exploitation would be inevitable. The beauty of forests, feeds the very souls of the prairie people:

> The aesthetic value (of the forest) cannot be over estimated. By the determined action of the earliest settlers who drove past three hundred miles of arable prairie to reach the forested hills, by the ever-increasing summer camps, an appreciation of the presence of the forest is amply demonstrated. Even now there are from fifty to one hundred people who spend the summer camping in Turtle Mountains. As the population increases, days of ease grow nearer, and the active booming age of realism gives way to a more human sentiment of appreciation, this number will grow larger. Here they send their children during the long Summer vacations, here

> they entertain their visitor, instead of remaining cooped up in the little sun-baked red-hot prairie towns, and here, on the timber reserve is the one spot, which all citizens of such favoured localities value most highly.[8]

At the conclusion of his academic work at the OAC in the spring of 1906, MacMillan graduated from the University of Toronto (the degree-granting university on behalf of the OAC) with the degree of Bachelor of Science in Agriculture with honours in biology. Again Drushka sums up the man and suggests that it wasn't so much that MacMillan chose forestry but that "the profession chose MacMillan":

> He was an intelligent, capable young man, attracted to the challenges offered by forestry, intrigued by its scientific aspects and attuned to the moralistic overtones of its conservationist ethic. MacMillan impressed established members of the profession, and they went out of their way to involve and court him.[9]

After graduation, he returned to Manitoba with the federal Forest Service and was posted to the Riding Mountain Forest Reserve. He had made up his mind – he would become a forester.

H.R. MacMillan in a dense forest of black spruce Riding Mountain, Manitoba, 1906.

Graduate studies

His application for the two-year Master's degree program in the Forestry School at Yale University in New Haven, Connecticut, was successful. At last, he was pursuing his passion for studies in forestry to which he gave all his enthusiasm and his energy and his professors described him as an avid and brilliant student. In fact, he did so well that he received permission to miss the third term. He left immediately for Powell River, British Columbia, where he served as a timber cruiser* and staked lease claims on timber stands for a syndicate of businessmen from Lindsay, Ontario. During the summer, he staked a claim for this employer which, years later, MacMillan himself would own within the MacMillan, Bloedel & Powell River Company.

MacMillan's summer work in 1907 provided him with an introduction to British Columbia, a province endowed with forests beyond his wildest imagination. Intrigued by the frontier, he visited and worked in many remote areas that deepened his concern about the working conditions of employees and the lifestyles of the people there. He took note of industrial development in the forests, particularly their exploitation, which showed little commitment to forest management and preservation. Entrepreneurs saw the forest industry as a "get-rich-quick opportunity" and took every advantage afforded them by the McBride government's offer of timber leases. More importantly, MacMillan was discovering the vast forest reserves of Canada's West Coast, and he loved what he saw despite the rampant exploitation.

A budding capitalist

In the fall, with his summer earnings in his pocket, MacMillan returned to Yale University and resumed his studies with devotion and enthusiasm. Elected vice-president of his class, he delivered the class address at the graduation ceremonies the following spring.

In his last term, he traveled to Wesgufka, Alabama, for his fieldwork assignment. Before he set out on the journey with only twenty-five dollars for the trip, his student colleagues made a wager with him – they bet him a like amount that he would not arrive in Alabama with all his money. To their surprise, he was able to double his dwindling cash

---

* A person who surveys forest land to locate timber and estimate its quality by species, size, quality, and potential products.

along the way. He described his experience in a letter to his cousin, Mazo:

> We were given about ten days from the close of the winter term until work opened up in Alabama. These ten days were no good to me as I had nothing to do and no place to go. (Willis Millar) was in about the same financial position as I, and we decided to work our way to Alabama. Some one said we could not do it. I had about $25 left, and I made a bet of $25 that we could do it. We did it, but it involved sleeping two nights in Central Park in New York City in the month of March, and going down from New York City to Charleston in the steerage of a boat ..., which was a cheap but long-remembered trip, and walking or beating our way on freight trains from Charleston to our final location.[10]

During the second day, he was joined by a dog. On the fourth day a man stopped him and said, "Fine dog you've got there, mister, want to sell it? Twenty-five dollars exactly," said H.R.M. Then he went to collect his winnings – or they may have been his earnings.[11]

Poverty had plagued his childhood and continued to do so through his graduate college years. The reality of this did not escape one of his fellow students:

> Money was still a grim and pressing problem to MacMillan. The little he had earned during the summer plus what his mother could send him was barely enough to meet his basic needs; there was none left over for frivolities. "My first impression of Mac," one Yale man said recently, "was that he was very poor. His thinness of body and his poor clothes showed that he didn't have much money to spend.[12]

In the spring of 1908, he kept his commitment and delivered the class speech at convocation. He graduated *cum laude* with the Master of Forestry degree. Although there were only a dozen jobs in North America, he probably could have chosen any one of them but he chose to return to Canada to become assistant inspector of forest reserves for the federal government. His monthly salary was one hundred dollars.

MacMillan's new work took him to Manitoba, to northern Saskatchewan north of Prince Albert, to the Waterton Lakes, and to Cardston, Alberta. His assignment was to survey these areas as a first step in the federal government's plan to establish national parks. Yet his summer was not all work – he took timeout to visit his Yale colleague and friend,

Yale Forestry School graduation class of 1908. MacMillan is leaning forward in the second row, third from the right. The photo was taken on the timberlands of Kaul Lumber Company, Wesgufka, Alabama during the spring term.

MacMillan was passionate about living in the outdoors and is seen above visiting Idaho, in October, 1908. A month later he was diagnosed with tuberculosis.

Willis Millar in Priest Lake, Idaho. Together they explored the backcountry of Kanisku National Park, which included a visit to a former Native campsite.

Drawn by the pull of the mountains and wilderness, he planted his spirit firmly in their vast tracts of forests. Although he had seen eastern and western Canada, the west, particularly British Columbia, caught his imagination and struck a chord within his innermost being – he fell in love with it and knew he wanted to stay. He wrote to Edna Mulloy, his fiancée, suggesting that on day they might live together in this part of Canada.

Chapter Three

A Time to Heal

A major turning point in MacMillan's life occurred during his survey work in Alberta. As a result of overexposure to the elements while camping in the remote mountains near Cardston, he developed a severe and persistent chest cold. In November, with his condition worsening, he returned to Ottawa to work for the federal Forest Service. His physical health continued to deteriorate seriously and he was diagnosed as having contracted tuberculosis. One characteristic of TB is that the disease's bacteria can lie dormant for years in the host's body. As a child, it is probable that MacMillan was infected by the same disease that killed his father and many close family members. Any antibodies he may have built up provided only limited protection against it, and he wondered if he would succumb to his family's disastrous history of illness and death.

In the early 20th century, little was known about TB except that most cases were usually fatal. MacMillan understood immediately how vulnerable he was to a similar fate. One small hope surfaced in the news that a medical practitioner called Dr. E.L. Trudeau, a TB survivor himself, had recently established a sanatorium at Saranac in the Adirondack Mountains in upstate New York, not far from the Canadian border.

Admitted there for what he thought would be a three-month treatment, MacMillan was shocked to learn from the attending physician that he had a one in ten chance of surviving. However, this was only a conditional prognosis, and his doctor told him he *might* survive if he meticulously observed every order and condition prescribed to him.

Young MacMillan quickly pieced together what he knew of his own father's approach to the disease, an approach that had probably accelerated his death. His father had succumbed to TB because he had kept working and had spurned rest and fresh air, the only known cure for the disease. This son was determined not to follow his father's folly:

> MacMillan spent all day in a steamer chair on an open porch at Saranac Lake, doing little except reading and thinking. There were a number of interesting characters in the sanatorium with him, from all over the world. There was the fellow writing articles for American magazines. Another, a mathematician who had been there for eleven years, spent his time working out astronomical calculations for the Canadian government.[1]

Published anonymously, MacMillan wrote of his experiences in the sanatorium.[2] His work was featured in *The Canadian Magazine,* under the title, "Fighting the White Plague":

> From other patients I learned how to be comfortable, how to avoid chills, how to keep warm in cold weather. The open porches in the winter ... are the same temperature as the outside. [I wore] ... a hockey cap, woolen lined leather mittens, felt boots, and over my ordinary clothing a fur coat and two heavy woolen blankets ... I had a hot soapstone at my feet.
>
> I expected to be able to walk around a little, play cards whenever I felt like it, write multitudes of letters, read, study and enjoy myself. I soon found from the doctor and other patients at my boarding house that I was far wrong, that I could do none of these things, [and] must not, for instance, climb stairs oftener than absolutely necessary, must take as few steps as possible and take these few slowly. I also learned in the first day or two that I must stay quiet in my chair eight or ten hours a day and spend the rest of the time in bed, except for the hour or so devoted to meals. And such meals! My appetite, which had been poor, increased within the first few days to such an extent that I found myself eating more than ever I had on hard exploration trips.[3]

It soon became painfully apparent that no cure would he realized for a least two, perhaps three, years. MacMillan remained at Saranac only briefly, until it began to look as though he was running out of money. The reality was that the cost of the treatment, thousand dollars per year, was more than he could afford.

Dr. Lorne McGibbon, a wealthy Montreal financier, had opened a Canadian sanatorium similar to the one in Saranac. For MacMillan, this was a providential event, and he began his long stay as one of eight patients admitted in the first boarding house "hospital" on the corner of

Rue Prefontaine and Rue Albert, Ste-Agathe-Des-Monts, in the Laurentian Mountains about sixty miles northwest of Montreal. Ste-Agathe is fifteen hundred feet above sea level and its air quality was described as "champagne." The annual $600 cost for treatment and lodging at the sanatorium was a more realistic one for the MacMillans. Initially, the Federal Department of the Interior had allowed H.R. to take only three months paid sick leave. Later, however, his employer sent him documents to edit or for analysis, all of which made him feel remembered and useful.

His mother, Joanna, was hit hard with the news of MacMillan's diagnosis. Her worst fears were unfolding before her – that her only child was now a consumptive and perhaps his fate was sealed. TB had plagued her throughout her life: while she was still a young mother, her husband had died of it, and her mother and her five brothers had died of this disease as well. Determined to help her son after he had been in the sanatorium for five months, Joanna left her job as a housekeeper near Aurora and moved to a small apartment in Ste-Agathe where she could care for him while he was receiving treatment at the sanatorium.

Charting his progress

MacMillan spent his time reading classical literature plus every book on Canadian history he could find. He decided to record the progress of his medical condition much as a nurse would:

> I got worse for about a year, at the end of which time it was apparently very doubtful which way I was going to travel. The outstanding event of that length of time from my standpoint was one grey day in the winter: I was lying in my room, which was a narrow one, upstairs, facing the northwest, from which the windows had been removed and through which I could see nothing but the grey sky, and, generally speaking, I was feeling very sorry for myself.
>
> The doctor came to see me. He had a talk with me and judging by my symptoms, of which he had been keeping a record, the time had come when I had to give the job all I had or it would be too late. I remember feeling very low. I had taken the cure with all the concentration and care of which I was capable.
>
> I remember that for months I kept a chart of what hour of the day I first coughed and how many times a day I coughed and every day I tried to delay the hour of the next cough and to reduce the number of coughs – I

very nearly suffocated myself in the process. Suddenly I began to recover very rapidly and at the end of fourteen months I was allowed five minutes exercise once a day, which consisted of walking ten or fifteen steps and back again, and for the next six months the exercise was increased until it reached an hour twice a day, which was practically all the freedom in the world.[4]

MacMillan did not recover as quickly as he had assumed or hoped he might. However, as he began his third year of treatment, his health improved and his energies began to return. He lived a solitary life, read constantly, and saw almost no one except his mother. His thoughts were never far from his fiancée, Edna Mulloy, which was especially true when his spirits were low or during the holiday seasons that celebrated family. On Christmas eve of 1910, he took pen in hand and wrote to his beloved:

Little one mine,

I cannot help writing to you again, even at the risk of making you weary of me. Oh but I have had the blues – yesterday and last night – so much so that last night I could hardly sleep. I am ashamed to tell you – and did not intend to – I am so afraid you will think it cowardice of me, foolishness or some such thing – to feel so downhearted. But I tried not to. I think probably it was chiefly loneliness and homesickness for home, homesickness to see you. I am terribly tired of this, tired to death of lying around when everyone else works and has a grand time, tired of the never ending uncertainty of it all.

If you had asked me in that letter I got yesterday, I would have gone to Aurora for Xmas. I am so anxious to see you, to get away – just as well you didn't, you will be having a grand time, and if I had been there I would have broken in upon it. Oh, I am better off this way, I know it myself, but it makes me sick to think it.

But it will be different another year. I feel sure that if I am careful we can get married this coming year and so long as I think that you may be sure I'll be careful. Otherwise, I think sometimes I would just let things go – this never ending strain, this constant stretch of uncertainty and monotony are too much.

I feel I must see you before long. I do hope I can. I always get so much comfort from your company and so much encouragement from your love. Oh dearie, you do love me, don't you? If only I could see you – if only I

> could hear you tell me so – if only I could feel the warmth of your caress, the pleasant intoxication of your presence – I would be so much better. A merry Xmas dearie. Yours,
>
> Reginald.[5]

Post-graduate course in TB

The lessons MacMillan learned while he was a tuberculosis patient in two hospitals served as professional courses in life skills. He had been a high-energy, work-driven, and rising star of the forestry profession who was now struck down and hospitalized. At the threshold of his career, he was no longer the master of his own fate. But forces were at play in his life – he needed to instill within himself new discipline to save his life. To do so, he would have to nurture his inner spiritual side. Later, he reflected on this experience:

> I took a post-graduate course in tuberculosis. It was a shocking thing. On the eve of getting to work, after getting through university, to find that I was sick. It was even worse to be told very soon that I might not get better. And that if I did get better it would be a matter of two or three years. It took some time to accommodate myself to that condition. I very soon discovered that there was a good deal of effort [needed], it had to be steady effort, and I had to change my whole philosophy. I couldn't think in short terms. I had to find some way of keeping my mind off myself, to become detached respecting the job I was on and I did it by reading, which fortunately I was interested in doing, and I read an average of two books every three days or so for two years and a half. Naturally that developed a steady habit and that was the most valuable part of my education, I think.[6]

The extent of MacMillan's reading of Canadian history was of doctoral-student proportions. During the months that he lay on his back in Ste-Agathe, he found his intellectual nourishment in the story of this young nation. The study of history fascinated him and proved to be a foundation for his future. As the years unfolded, he assembled one of the finest personal libraries of Canadian history to be found anywhere in the country. According to Roland Wild, "Few Canadians can approach his detailed knowledge of the country down from French colonial days."[7]

An articulate disciple

It was during his convalescence and his long months and years of reading and contemplating the surrounding woods that MacMillan embarked on a new and revolutionary way of understanding human interaction with forests. Drawing on his academic and professional training, he began to document his ideas – his conservationist ideals and his thesis on the need to prevent forest fires. Gradually his thoughts assumed a decidedly political dimension, and he wrote arguments against the unrestricted export of forest resources. Remarkably, his written work anticipated the conservation and environmental movements that arose in the latter part of the 20th century and came to the fore in the 21st century.

MacMillan's conservationist ideas are best understood in their historical context. Until this time, the conventional view of forests, particularly those bordering the western prairies, was that they were impediments to the settlement and growth of society in North America. Timber resources were viewed in much the same way as mineral deposits: their supply was unlimited and they were there to be extracted. Once the forests were cut, the land would be suitable for settlement and agriculture and any land not used for these purposes was considered to be "waste."

In sharp contrast, MacMillan's conservationist position suggested that forests should be permanently valued and maintained, and he argued that a timber economy could be based and built upon on this premise. Furthermore, he believed forests should be harvested but only on the condition that they be allowed to regenerate naturally. The reality is, however, that the way in which forests are harvested, that is the methods used for extracting their resources, determines how well nature can renew itself *without* human intervention. MacMillan was a young and articulate disciple of the conservationist position, but he was concerned not so much with forest preservation as with the well-managed use of forests. The process of harvesting, he argued, is directly related to the forest's own capacity to regenerate naturally. The destructive practice of "clearcut" extraction that was widely used in the latter half of the 20th century has dramatically illustrated the importance of MacMillan's insights.

Finally liberated

By the summer of 1911, after thirty months of treatment, MacMillan was well enough to leave the sanatorium but only with the promise that he would keep up his strict medical regimen of resting during the day. In his letters to Edna, he shared that he felt very sorry for himself. Despite these feelings, there is no suggestion that he remained bitter about what others might perceive to be a waste of almost three years of his life in a sanatorium. There is no record of a Job-like internal debate or any raging against God for his fate. Indeed, MacMillan's feelings were quite to the contrary: his experience in the sanatorium fighting his disease had provided him with lessons for life, for survival. He had used this time to regain his health and to read constantly in preparation for the future. Rooted and grounded by their love, his family provided him with every care. His spirit remained positive, and he felt he would be able to regain and maintain his health.

Upon his discharge from the sanatorium, MacMillan returned to his work with the Forestry Branch in Ottawa, and his mother returned to Newmarket.

On August 2, 1911, fifty family and friends gathered to celebrate the long-awaited marriage of Harvey Reginald MacMillan to Edna Mulloy. Their wedding took place in St. Andrew's Presbyterian Church in Aurora, and Edna's two sisters, Flossie and Olive, were her bridesmaids her while the Reverend James Strachan served as MacMillan's groomsman.[8]

With marriage, MacMillan's formative years were now complete, and his professional career was finally beginning. This period was one of rapid economic growth in Canada, which the *Canadian Courier*, the country's first newsmagazine, summed up with the capitalistic optimism of the new era in 1909:

> ... hundreds and thousands of Canadian boys can hope to trip over a fortune one of these days in the Canadian wilderness. Pots of gold lie hidden there more truly than at the end of rainbows ... Canada is a perpetual Christmas Tree with a present for every son of the house.
>
> We are adventurers living the thrilling life of discovery, of daring, of chance and of conquest. Bare-footed boys become millionaires and live in palaces. In the knapsack of every Canadian schoolboy there is – not a marshal's baton – but a millionaire's bankbook.[9]

St. Andrew's Presbyterian Church, Aurora, Ontario where H.R. MacMillan and Edna Mulloy were married in August 1911.
Credit: R.D. MacRae based on the sketch of Jim Watson, Aurora.

MacMillan, with his forestry degree and years of reflection behind him, was ready to seize the opportunities that lay before him in the west, and it was not long before the west took notice of him.

Chapter Four

A Time to Mature

For half a century, H.R. MacMillan's rise through the forest industry was by any standard, meteoric, sudden, and brilliant. His preoccupations from 1912 through to the 1930s were to implement his ideals for forest management: to establish his own business, to shape its infrastructure, to build an industrial base, and to seek worldwide markets for his products – lumber, newsprint, and fine paper. Due to his systematic approach, his clear definition of goals and objectives, his knowledge of his product, and his acute sensitivity in building a network of contacts, MacMillan's rise to a position of power and influence seemed predestined. His influence and financial resources grew to enormous proportions in relatively few years. The *Financial Times* of April 1, 1938 described MacMillan as "the No. 1 industrialist and business leader of British Columbia."[1]

Whether or not his stature in British Columbia's business community was fully realized, MacMillan's success had come about because of his considerable personal talents, his seemingly inexhaustible drive and energy, and his deep and instinctive insight into human nature.

MacMillan was a man familiar with world history who must have realized he had become a Caesar-like figure of the forest industry. Like Julius Caesar some twenty centuries before him, MacMillan may have mused on the following personal assessment: "*Veni, vidi, vici,*" said Caesar, "I came, I saw, I conquered."[2] By the late 1930s, MacMillan was at the top of British Columbia's forest industry and fifteen years later he towered over all others in Canada.

Disciplined approach

One of the secrets of MacMillan's success was his basic attitude toward life and learning. He viewed every door that opened to him as a God-given opportunity. His was an instinct born of poverty, formed in faith, and nurtured at Ste-Agathe, and the lessons he learned on survival stayed with him for life. The first and most important lesson was to care for his own health. To this end, and despite his vigorous work habits, he followed a strict health regimen that took special account of his need for rest. A second lesson and his attitude toward life was to embrace opportunity and set impossible goals for himself. Far too often, he believed, people confined themselves to targets they feel certain to achieve. As far as he was concerned, this attitude was a self-limiting one. A third lesson, and one that he followed to the end of his life, was that there were always more things to learn and more yet to accomplish.

As MacMillan plunged into the new challenges before him, he knew that some might demand more than his energies and intellect could match, but he was not intimidated. One memorable story that illustrates this quality involved him chastising and firing Henry Pybus Bell-Irving (who would one day become Lieutenant Governor of British Columbia) because he did not meet MacMillan's expectations on the job. After Bell-Irving pleaded the circumstances leading to his mistake, claiming that it was impossible for him to have done otherwise in the exercise of his duties, MacMillan retorted in imperial and towering tones, "Young man! You must understand that nothing is impossible!"[3] After some consideration, however, he rehired Bell-Irving. MacMillan's admonition stuck with Bell-Irving, who translated his exhortation into French, *Rien n'est impossible,* and registered it as his family motto.

One cannot help but wonder if that stern admonition to Bell-Irving came from MacMillan's close acquaintance with the Holy Scriptures. He had, after all, led a Bible study group at university. "With God," wrote Luke and Matthew in their gospels, "nothing will be impossible." "If you have faith as a grain of mustard seed," said Jesus of Nazareth, "you will say to this mountain, 'Move from here to there,' and it will move, and nothing will be impossible to you."[4] Both passages would have been familiar to, and easily recalled by, MacMillan.

Chief of the woods

His rise to becoming the Caesar of British Columbia's forest industry began in July 1912, when he accepted the newly created position of chief forester of the province with an annual salary of three thousand dollars. It was a time of immense economic growth. Construction was at an all-time high. Years later, in 1969, he would unveil a bronze plaque at the opening of the new UBC forestry-agriculture building named in his honour. In his remarks on this occasion he modestly referred to his own appointment as chief forester at the age of twenty-six years:

> "As it happened, when I got there," MacMillan said in the tongue-in-cheek manner he sometimes used in recalling his past, "they thought they'd hired the wrong fellow. I was too young. But I fitted the specifications that they had to have a trained forester and he had to be a Canadian. It happened in that particular year, 1912, I was the only one that could say the specifications fitted me. I got the job, but when they looked at me, they wished to God they hadn't got me. I looked too young and fragile and unwise. But the Minister was too proud to fire me. He waited. I think that's a habit of Ministers. He waited until I had made a mistake, and I made a good many, but he didn't catch me at them."[5]

In the four years MacMillan served as chief forester, he built one of the most important departments of the government of British Columbia. In doing so, he gained intimate knowledge of every aspect of BC's forest resource: its timber reserves, its manufacturing process from raw resource to finished products, and its marketing and delivery around the world. More than any other person, he knew the industrial potential of the province. This information and his thoroughly systematic and disciplined approach were, in no small measure, the keys to his subsequent success as an industrialist who would market his products worldwide.

The forests of British Columbia were relatively unexploited during the 19th century but rapid expansion in early 20th century began to wreak havoc on them. By the time MacMillan became chief forester, the challenge to bring order to rampant and uncontrolled industrial exploitation was immense, and he had the awesome assignment of putting in place the principles of good forestry management and practices. But he was able to do so in the face of economic and business pressures and in the context of British Columbia's politically volatile government.

H.R. MacMillan, the first chief forester of British Columbia, 1912.

MacMillan felt it important to keep the high standards and the integrity of his profession. He understood the larger picture of both the economic potential and the vulnerability of the forests to unbridled expansion and exploitation. He therefore felt it important to hire the brightest and most professionally qualified staff he was able to find and with these human resources in place, he began to apply his vision.

Trained foresters for all of British Columbia's seventeen forestry districts were not easy to find. In his search, he had the advantage of a buoyant economy and a political machine that accepted and endorsed his vision. His personal charisma attracted well-qualified people. At the very beginning of his appointment he told a Victoria *Daily Colonist* reporter that, "wisely and properly handled, British Columbia can grow timber and ship the products of the forest in perpetuity."[6]

Hiring the right people

The people MacMillan hired were leaders in their field. He drew foresters from federal and provincial positions across the country and despite considerable opposition from the United States. Many of his staff were to play significant roles in the unfolding forestry drama through the decades that lay ahead. Perhaps none became more significant in his life, and to the success of his career, than Whitford Julian VanDusen.

In the winter of 1908, it was Dr. Bernard E. Fernow, the new dean of the Faculty of Forestry at the University of Toronto, who introduced VanDusen to MacMillan. Their meeting proved to be another providential moment. From that introduction came a lasting relationship as lifelong business partners, and their friendship grew from there.

VanDusen was a bright student intent on a career in electrical engineering and he and MacMillan met for one hour in Toronto that day in 1908. So convincing and persuasive was MacMillan about the future of the forest industry in British Columbia, that VanDusen switched disciplines. After VanDusen's first year in forestry, MacMillan hired him to work as part of a survey crew in Manitoba and, in January 1913, appointed him assistant chief of management of the BC Forest Service. His initial responsibility was to develop the British Columbia export trade.

At this time, MacMillan oversaw twenty-six professional staff in his BC Forest Branch, who he had assembled from across Canada and the United States to serve in this rapidly growing and vital provincial department. In a photograph of the BC Forest Branch taken in April of that year, a youthful looking VanDusen stands in the back row while MacMillan, front row centre, towers over all his staff. In this first photograph of the BC Forest Branch, both he and VanDusen appear much younger than the other department members.

MacMillan's administrative and intuitive skills combined to make his first annual report an impressive and farsighted one. Once his staff was in place, he announced the directions the Forest Service would take.

The British Columbia Forest Branch staff 1913. MacMillan is center front row with light coat and VanDusen is third from left in the back row. Both men appear younger than all other members of the Branch.

> British Columbia contains not less than one hundred million acres of forest land, and the total stand of commercial timber is certainly not less than three hundred billion feet, and probably much more. At present the lumber output of British Columbia is about a billion and a quarter feet a year. At the present rate of cutting, making no allowance at all for annual growth, it would take the lumber industry of British Columbia nearly 250 years to use up merely the mature timber now standing. The annual growth of the forests of British Columbia is even now, before they are either adequately protected from fire or from waste, certainly not less than five times the present annual lumber cut ...
>
> What is not cut is wasted in the end. It is not merely advisable to encourage the growth of our lumber industry until it equals the production of our forests – it is our clear duty to do so, in order that timber which otherwise will soon rot on the ground may furnish the basis for industry, for reasonable profits to operator and Government, for homebuilding, and, in the last analysis, for the growth of British Columbia.[7]

MacMillan's 1913 report provided the basis on which the policies of the BC Forest Service were established for the next seventy or eighty years. He was not a provincial politician rallying support but a young and highly respected professional civil servant. What he had articulated became the rationale for promoting the growth of the forest industry as well as its management. Further, his report furnished the arguments politicians needed to endorse the rapidly growing infrastructure of the Forest Service.

Conservation a prime concern

The Forest Service's annual report was a distillation of MacMillan's thinking from his earlier academic days. His proposed policies reiterated the idealistic theories of his tutors and contained his own observations on the dangers threatening the forests of the western provinces of Canada. His annual report attempted to mediate the ethics of those who would exploit and destroy the forests and those who argued that timber should not be harvested at all. The years 1912 through 1915 were a time of a rapidly expanding economy and forestry was fueling the growth. MacMillan set the parameters of the Forest Branch to control the boundaries within which the forestry industry could operate. His policies promoted responsible forestry practices *in order to* protect forest resources from those who would irresponsibly harvest the woods by destroying and abandoning them. He felt strongly that the only way to protect the forests was to develop an industry that valued and cared for them in perpetuity.

MacMillan's belief in the importance of the stewardship of the earth – an expression of his personal theology – was rooted in a biblical understanding of creation wherein the divine hand cares for it. This belief reveals a private and spiritual dimension to MacMillan that is not generally recognized. An address he delivered that focused on the care of forests was quoted in the Victoria *Daily Colonist*:

> Denude the hillsides, either by fire or by reckless, *improvident*[*] cutting, and when the rains come they wash all the earth from the mountainsides, flood the valleys and in time carry off all the earth there, leaving a treeless, desolate waste incapable of growing a blade of grass or supporting the least of living things.[8]

The "hot" issue of conservationist versus industrialist that MacMillan's 1913 report provoked clearly bothered him – his scholarly

---

* Author's emphasis.

integrity was being challenged and the principles for which he stood, and his newly launched profession itself, were at stake. Biting criticism came from industrialists who had enjoyed unbridled access to forest resources. They understood MacMillan's new policies as a rein on their profits and they did not like it. He felt keenly the antagonism expressed toward the Forest Service policy he had established and he took it personally. In sum, he believed the anger directed toward him was based on "misinterpretation, misconception and misinformation"[9] and he set out to put the record straight – managed forests can also be profitable for business interests. In Ottawa, the following year, he spoke to the annual meeting of the Conservation Commission and argued:

> The future of this country depends upon our making every acre productive. Broadly speaking, the earth's surface can be made productive in two ways only, by producing agriculture or timber crops ... Optimistic as we have been in this country, we seem to have been unable to see any value or worth in caring for our non-agricultural lands ... We must realize that our present stands of merchantable timber cannot support our growing industries indefinitely ... The future forest industries ... must be supported by the timber grown on the logged-over and burned-over non-agricultural lands. Looking at these lands we should see not wastes holding no promise for the future, but productive lands needing only protection from fire to enable them to support logging camps, pulp mills, rural, and industrial communities of a type which has done much for Canada.[10]

What he had articulated so well in his address, and a point that was the highlight of the Commission's meetings, was that Canada was doing little for the protection of forests. The policies he proposed could be mutually beneficial to both the lumber industry and the conservationists. Senator William Edwards, impressed by MacMillan's speech, suggested that one day he might become the prime minister of Canada.[11]

Special trade commissioner

By 1914, ominous events were unfolding on the international stage. It was an uneasy time for Canada and for British Columbia as the threat of war in Europe persisted. In addition to economic and political uncertainties, the declaration of war had an almost immediate impact on MacMillan and the Forest Service itself. The fervent patriotism his staff felt toward nation and empire meant that this department was soon faced with the departure of its staff who went to fight in World War I.

In March 1915, MacMillan accepted an invitation of the Government of Canada to serve as its special trade commissioner to seek world markets for Canadian lumber in Europe, Australia, Asia, and South America. When his new appointment and world-trade tour was announced, it generated great interest in the media. Expectations ran high in British Columbia with headlines in the daily newspapers such as "MacMillan Ready to Leave for Australia," "Chief Forester to Circle the Globe," "MacMillan Goes to the War," "Trade Commissioner to Leave Shortly." San Francisco shipping companies, on the other hand, received the news with alarm and immediately announced a fifty-percent increase in shipping rates between BC and Australia.[12]

The trade mission took MacMillan around the world to meet government and business leaders connected to the forest and lumber industry. He and his wife left for London in April, travelling by rail across the US and then by ship to the UK. Upon arrival in London, MacMillan closed his first deal in just two weeks: an order for ten million board feet of BC lumber from the British Admiralty. He now had the proof for what he had argued publicly – namely, that the hope for BC's forest market was to be found overseas. One huge obstacle, however, was that the United States controlled world markets and gave favourable treatment to their own producers. MacMillan described the assignment:

> "I went to the UK in April 1915," MacMillan said, "to represent the Canadian government at a time when export trade to the UK from Canada had been cut off. My job was to get the export [market] for Douglas fir from British Columbia, and spruce and pine from eastern Canada, started as soon as possible and get it into Canadian hands."[13]

By the end of July 1915, he had orders for a further twenty million board feet of lumber. BC newspapers echoed the ecstasy of both government and industry at the success of the MacMillan world-trade tour, especially because the new contracts bypassed third-party American agents who had previously controlled the marketing of Canadian timber.

Worldwide markets

Opportunity continued to shine on MacMillan. Sir Richard McBride, the former premier of the province of British Columbia, was now BC's agent general in London. Together they convinced the UK Board of Trade to purchase lumber from Canada rather than the United States in order to bypass the interests of third-party agents. Britain was at war, and the US was not, and the efficient delivery of lumber products was crucial to the war effort. Not the least of which was Sitka spruce that the UK government needed for aircraft construction.

It did not take long for MacMillan to appreciate that no commercial organization in BC could cope with the contracts he had obtained. From London, he ordered the BC Forest Service to arrange for delivery services to be organized under the direction of his friend and confidant, Julien VanDusen. This had the immediate impact of convincing private mills to act together in supportive response to overseas orders that kept pouring in.

MacMillan held the position of special trade commissioner for eighteen months. During this time, he visited potential markets in Britain, France, Holland, South Africa, India, Australia, and New Zealand. He left these countries with orders for tens of millions of board feet of lumber and with many personal contacts in governments and in industry that would ultimately become crucial elements in his future career. During these trips, he was introduced firsthand to the politics and economics of international trade and commerce. During his tenure as Canada's special trade commissioner, he addressed many groups of financiers and significant decision-makers around the globe, and he became known as one of the most knowledgeable foresters in the world. Ever mindful of the ability of his completed work to create new openings and to generate invitations, MacMillan later called his appointment as Trade Commissioner, "one of the great opportunities of my life." In this role, he proved that Canada was able to market its timber without American advocacy.

The apparent ease with which he obtained contracts for lumber orders no doubt contributed to his growing desire to go into private business for himself. The elements of his future in private business were beginning to come to the fore – his professional expertise, clear business sense, and analytical skills; his staff-management and problem-solving abilities; and his solid and widespread network of business connections.

Resignation as chief forester

Upon his return to Canada, MacMillan discovered his Forestry Branch had been virtually emptied by attrition. Many of his professional foresters had enlisted in World War I and were not replaced; others simply did not return from the war. This loss of staff, together with his growing desire to enter private enterprise, led MacMillan to resign as chief of the forest service.

There was an element of high drama in his announcement. MacMillan owned a limited-edition book of secret codes used for telegraphic communications which, along with other items of personal interest, he kept in the drawer of a side table in his office. These codes consisted of a succession of numbers cross-referenced to words that were used to transmit secret messages.[14] For the purposes of his resignation telegram to The Honourable W.E. Ross, B.C.'s minister of lands, however, he chose not to use these codes. MacMillan devised his own code, presumably because the published book of codes, despite its limited circulation, was available to others. Although there was a remote possibility that someone other than its intended recipient might decipher his letter of resignation, he took precautions to prevent anyone other than Ross from understanding its politically sensitive message. He turned to the Bible for the words used in his private code: names of books from Hebrew literature; selected terms from both the Old and the New Testaments; and words associated with the early Christian community.[15] His familiarity with biblical literature, not far removed from his daily life, was clearly present in this secret code, a system carefully worked out between MacMillan and W.R. Ross. Code breakers attempting to crack open the system, and therefore his message, would have found it to be enigmatic:

> AN TELEGRAPH VARIED PROLIFIGATELY YEAST WREATH OBLIGED INNOXIOUS HOSEA OPPIDAN NEAREST RETAIN THROBBING HOSEA FORMULA ISRAEL PHARISAICAL SWELLED EFFACED EZRAS DIGRESSING OPPIDAD FRACTION BRIDGE ANTARTIC SUCH REDDISH CONNECT GRAINGERS PRIMED ARCHFIEND ASSESSOR CHUB FRACTURE DRAFTED OAKEN DEUTERONOMY PUP ANTIQUITY NEAREST RETAINED BENEDICTINE CONCERTEDLY NEAREST DRIER YEAST NOOK WIDE UNTIL HOSEA TILLED NIBBLINGLY MACMILLAN[16]

MacMillan cabled his resignation message* to Ross, in Prince George on September 12, 1916. He told the minister of his intention to take up private work, and he was presumptuous enough to recommend his successor to Ross.

In only four years as British Columbia's chief forester, MacMillan gained wide support for his massive contributions to the industry which profoundly changed the economy of British Columbia. The Vancouver *Daily News-Advertiser,* lamenting that the government had now lost his service, opined in an editorial:

> He will still be in British Columbia, associated with the management of one of the largest lumber concerns in the province. He will still be interested in the preservation and restoration of forests, and in the discovery and development of lumber markets. But henceforward the private business, with which he is connected, instead of the public, will have the first claim on his time and thought.[17]

Invitations from universities

Universities in the east had noticed MacMillan's ascendancy to significant global leadership in the field of forestry – universities in the United States, as well as the University of Toronto, offered him appointments, but he declined their invitations. In the west, Dr. Frank Wesbrook, the first president of the University of British Columbia and a close friend of MacMillan's, invited him to head UBC's new Department of Forestry and again, MacMillan declined. Rather than accepting any invitation to join an academic institution, he decided to work in the industry in order to gain hands-on experience in lumber manufacturing. He wanted to learn the essentials of the business world from the ground up, yet his first venture into the industry was not entirely positive.

---

* Translation:

AM TAKING UP PRIVATE WORK WISH NOW INFORM YOU OF MY RESIGNATION THANK YOU FOR YOUR PERSONAL SUPPORT DURING YEARS DEVELOPMENT OF FOREST BRANCH AND STRONG RECOMMEND CONFIRMATION GRAINGER PRESET APPOINTMENT AS CHIEF FORESTER DO NOT DESIRE PUBLIC ANNOUNCEMENT MY RESIGNATION BEFORE COMPLETION MY DOMINION WORK NEXT WEEK UNLESS YOU THINK NECESSARY. MACMILLAN.

"I'll own this outfit"

When E.J. Palmer, general manager of the Victoria Lumber and Manufacturing Company in Chemainus on Vancouver Island, invited him to become the assistant manager there, MacMillan accepted. His responsibilities were extensive and exhausting, as author Ken Drushka describes:

> MacMillan's duties encompassed every aspect of the operation, from the woods to the loading wharves. The logging operations were completely under his direction, and within weeks he was overseeing the building of one new railway and rebuilding another, both of them several miles from the mill in opposite directions. He also served as general superintendent of the mill, the shipping dock and lumberyard, as well as sales manager. "I was expected to put in a full seven-day week. I took inventory on Sundays in the inventory season, finished it on New Year's Day. No overtime was ever paid to anyone.[18]

In September 1916, MacMillan and his family moved into an old house a few yards[19] from the historic 1860s water wheel in Chemainus, a replica of which still stands in this friendly community.[20] Both he and Edna loved Chemainus. However, working seven days each week proved far too demanding on his health and, later that year, he was re-hospitalized for a short period. By contrast, Palmer, the general manager, rarely came out of his office to work in the real world of the lumber industry, and his main preoccupation seemed to be making arrangements to play golf on long weekends in Victoria. MacMillan's brief hospital stay was sufficiently ominous for him that he quickly sized up his ailing health and its possible implications for the future. Feeling over-extended and exploited on the job, it is not surprising that MacMillan was soon in conflict with Palmer over authority, planning, and decision-making issues. His resignation in July 1917 prompted the president of the company, Tom Humbird, to arrive from Idaho in an attempt to mediate the situation. His attempt failed, and MacMillan agreed to leave the company in September. As he turned to leave the office for the last time, his final comment was prophetic, "When I come through this door again, Mr. Humbird, I'll own this [blankety-blank] outfit."[21] He later reflected:

Tom Humbird (left), owner of the Victoria Lumber and Manufacturing Company, and H.R. MacMillan with one of his ships and the Vancouver harbour in background. They are depicted in the street art of the lumber and mills town of Chemainus, BC.
Credit: Street painting by Constance Greig-Manning.

MacMillan stormed out of the door (on the right) of Victoria Lumber and Manufacturing Company, Chemainus in 1917. He vowed to own the operation when he came through it again. He kept his threat and had the door beautifully restored. Later, the company constructed a like door, on the left. It is currently used for general access to the office.

> "I just made it ahead of getting fired. I got the idea and I quit. I disagreed with the fella that had all the authority and I observed how this was being taken, so I quit. Otherwise, I'm sure I would have been fired."[22]

Before, during, and after his time with the Victoria Lumber and Manufacturing Company, he continued to be courted repeatedly for academic positions. He seemed to be under constant pressure to consider appointments at both the University of Toronto and at the University of British Columbia. His interest in the creation of a possible School of Forestry at the University of British Columbia remained constant and, although tempted to take a position there, he was not prepared to be its first dean.

Hands-on management

In mid-1917, war was still being waged in Europe and MacMillan felt it his duty to contribute to the war effort in some way. Within a month of leaving Victoria Lumber and Manufacturing, he was invited to become the assistant director of aeronautical supplies for the Imperial Munitions Board. He accepted, and he and Edna moved to Vancouver with their two and daughters, Marion and Jean. His one specific and urgent assignment to assist the war effort was to procure Sitka spruce used for the building of aircraft. In a few brief months, he negotiated some three hundred contracts to bring Sitka logs from the Queen Charlotte Islands, also known as Haida Gwaii, to the mills of southern British Columbia where they would be used to make made products in support of the war effort. It was a stunning achievement.

Among the people he recruited to undertake this critically important assignment was Julien VanDusen whom he had lured from the BC Forest Service. He appointed VanDusen superintendent of operations and stationed him at Port Clements on Graham Island.

Government leaders, at least those who had some appreciation of what was involved in this program, were deeply impressed by the quality of the Sitka spruce and the speed with which it was delivered to the aircraft-building industry. Perhaps few would appreciate, however, that this mission could only be accomplished by the powerful hands-on attitude MacMillan brought to every engagement. He was determined, focused, and efficient. A good example of his work ethic is found in his rigorous approach to the delivery of forest products. In order to achieve the speedy turnaround time of ships delivering logs, MacMillan did not

MacMillan and Whitford Julian VanDusen held a life long partnership in the public civil service, in business, and in friendship are seen aboard a launch of the Forest Branch of BC in 1918.

allow crews to leave their tugboats during the industry's peak season when logs were being towed from Haida Gwaii to the mainland. No shore leave was granted to crew members who might be tempted to celebrate in a local pub because MacMillan, ever the realist, knew there was always the possibility they might not return in time for their ship's next sailing:

> One night in Prince Rupert, a tugboat crewman slipped over the side and swam to shore for a night of carousing. The captain sent word to MacMillan, who was in town, that he would stay put until the crew was brought up to full complement. This ultimatum sent MacMillan out onto the streets of Prince Rupert, scouring the bars, restaurants and hotels looking for someone to fill the position. But in the late stages of the war, unemployed men were scarce in the port city. Finally, in desperation, he went to the local jail and asked the chief of police if he had anyone who could be "borrowed" for a couple of weeks. The chief checked his inmates and produced one who was dead drunk and snoring peacefully. Undaunted, MacMillan dragged him down to the dock, dumped him in a skiff and rowed out to the tug anchored in the harbour. When the crew hoisted the drunk up on deck they discovered he was their missing sailor. Marveling at how efficiently MacMillan got things done they hauled up their anchor and headed back to the Charlottes for another boom of spruce.[23]

Not only did government officials, academics, business leaders, and tugboat captains admire MacMillan's ability to get the job done, loggers in the forests and at the mill sites held him in the highest regard. They knew that he was primarily responsible for the tremendous wartime achievement of exceeding production expectations. Their respect for him was founded on the fact that he knew the loggers well: he had chosen many of them for the job; he knew every manager; he visited every logging camp; and he spoke with the lumberjacks where they worked. His presence, intimacy with each operation, and personal concern for workers and management made each of them feel as though he was part of the team. He expected their work to be done well, and they met his expectations.

At the conclusion of World War I, MacMillan was again invited to consider heading the School of Forestry at the University of Toronto. Again, he declined because he no longer wanted to work for lumber companies or government agencies. Other challenges seemed to appeal to him more: he was eager to establish his own business and to own a part of the forestry industry. While he was in London meeting with industrialists and financiers, an opportunity arose that was too good to pass up. He had a plan in mind, and he returned to Canada to implement it.

Chapter Five

A Time to Invest

In 1919, Montague Meyer, a British financier and industrialist who was intimately involved in the Sitka spruce operation during World War I, came to Vancouver to discuss a proposed business partnership with MacMillan. They had met four years earlier, in 1915, when Meyer was the official timber buyer for the British government and MacMillan was Canada's special trade commissioner. They now met to lay post-war plans, which included the establishment of a lumber export company with offices in Vancouver. Post-war reconstruction meant the demand for lumber was high, and they realized a business opportunity lay before them. The plan was simple: MacMillan would provide BC timber, and Meyer would sell it in the United Kingdom. It was to be an equal partnership, and MacMillan borrowed his share of the required investment. For collateral, he mortgaged everything he owned, even his house located at 1326 West 13th Avenue in Vancouver, along with his furniture. Although Meyer, acting under the name "Canusa Trading Company," suggested that he would cover the total investment, MacMillan insisted that each partner contribute the necessary $10,000. The Canusa Trading Company was incorporated in British Columbia in July 1919, which marked MacMillan's rise to the top of the BC business community. "I admire British merchant traders,"[1] said MacMillan, and the respect was mutual: a photograph of MacMillan sitting on Meyer's desk hung in the latter's office for years.[2]

MacMillan's export business was an immediate financial success and orders for lumber flooded into the Vancouver office. His marketing ground rule was: *Any amount above 20,000 board feet to any retailer anywhere.* Before six months had passed, MacMillan needed a manager, and he wanted none other than his long-time business colleague and friend,

Montague Meyer, a British financier, joined business forces with MacMillan in 1919. The partnership set MacMillan on his business career.

Julien VanDusen. The new manager was to own seventeen percent of the new company. By the end of the first year, the company declared a staggering profit of $250,000 on the $20,000 original investment, based for the most part on the sale of railway ties to the UK and India. Clearly, this business partnership and its subsequent activities were among the most important in MacMillan's life. To the end of both their lives, the relationship between him and VanDusen proved to be very complementary.

Many years later, MacMillan reflected on his investment in the business and industrial sectors of British Columbia. "All I set out to do was to make a living," he said, "for, at the age of thirty-four at the end of World War I, I found myself without a job."[3] In fact, he set out to put British Columbia timber on the global market map through the same export strategies he had suggested to the Canadian government much earlier. He knew opportunity had presented itself to him once again. "The times," he said, "were with us."[4]

Vertical integration

The story of the creation of MacMillan's business empire, is both complex and intricate. Local and national building booms, along with economic depressions, directly affected forest interests, and MacMillan and VanDusen were on a persistent search for markets. From the start of their business relationship, either of them was constantly travelling the world in search of customers. During a twenty-five-year period, eighteen company representatives visited twenty-three countries other than the US, for a total of one hundred and six trips.[5]

The growth of the business empire had its genesis in many "seedling" operations, as MacMillan called them, which were tentative steps towards investment in a variety of aspects of the forest/timber industry: saw mills, plywood, and pulp mills. In June 1924, the company continued its expansion by establishing the ship-leasing Canadian Transport

H.R. MacMillan, a senior business man.

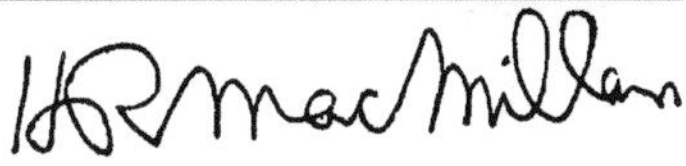

Company. Its distinctive corporate flag, a green Douglas fir on a white background, flew on ships entering harbours all over the world. MacMillan's shipping and transportation needs were accommodated at highly competitive rates. Furthermore, he established business offices in Vancouver, Montreal, Seattle, Portland, Tokyo, Yokohama, Kobe, Sydney, Shanghai, New York, and Melbourne. The vertical integration of the forest industry – that is the ownership of all processes involved in the planting of a seedling to the delivery of finished lumber products – was his operative business principle. His expanding interests and invest-

MacMillan meets with colleagues at his Franklin River operation on Vancouver Island in 1957. He was a hands-on manager of an industrial empire that merged with the forest interests of Prentice Bloedel. They discuss the operation with skidder foreman Bert Harris (center).

ments included everything from the harvesting of trees and the preparation of lumber products, to marketing and shipping them around the world. In August 1951, with the merger of H.R. MacMillan Export Company and Bloedel, Stewart and Welch the new forest company, MacMillan Bloedel Limited, became the largest in Canada. MacMillan held the controlling interests in the new venture.

Food on the table

As MacMillan's financial interests grew, so too did the number of people working for his companies. In the 1940s, MacMillan had 5,000 employees on his payroll. From time to time, there was talk among union leaders of possible strikes and open-shop union contracts cov-

When Prentice Bloedel merged his company with MacMillan's interests in 1951, they formed the largest forest company in Canada, MacMillan Bloedel Limited.

ered all hourly employees. However, MacMillan did not worry about these threats because he knew the men themselves agreed that they never had it better. One union official said, "H.R.M. seems to have been one jump ahead of us again; the conditions are so good, and the old silver-lip has his ear so close to the ground that we would have difficulty in finding a reason for improving our condition."[6] They knew, for example, that the company's safety drive had cut the accident rate far below average for the forest industry. Furthermore, the tables in lumber camp dining halls groaned under the weight of generous offerings of food, including roast beef, pies, and fruit. The loggers paid only half the cost of the meal and food was clearly important to the men in the forests as well as those serving on tugboats. Like Napoleon's military proverb, the "army" of forest workers in MacMillan's camps "marched on their stomachs" as well. Not surprisingly, "MacMillan" and "H.R." became household names on the BC coast.

Many years later, MacMillan recorded some of his thoughts and experiences about the elements in his business career that led to his success. He thought this information would be helpful to Mr. Justice J.V. Clyne, the man whom he chose in 1958 as the new chair of the board of directors of MacMillan Bloedel.

Clyne had little business experience at the time and MacMillan's advice was clear, "The top men, including the chairman, the president,

MacMillan enjoys a cartoon presented to him by W.D. Bill Moore of Winter Harbour, BC at a 1965 convention of the BC Truck Loggers' Association in Vancouver. The cartoon depicts MacMillan when he became BC's chief forester in 1912.
Photo credit: B.C. Jennings, Vancouver.

as well as responsible VP's," he instructed, "should show up around the operations without laying on a 'do,' and without any of the 'drill' as in a super-organized institution like an army. The local top man is sufficient to meet and go around with. Staff," he said, "is the brain, heart, circulatory, and nervous system of a business."[7] In his eleven-page paper entitled *Looking Back on This Business,* MacMillan laid out his management style and his advice on conducting business. "M&B must be maintained [as] a high-ball outfit," he wrote for Clyne, "doing the best possible [job] for shareholders, employees, its communities."[8] In his view, all three groups deserved the best and most farsighted leadership.

J.V. Clyne succeeded MacMillan as chair of MacMillan Bloedel in 1958.

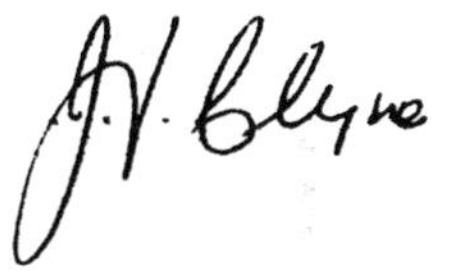

In addition to his forestry interests, MacMillan had other investments. For example, in 1930 he became a director of BC Packers, the largest fish-processing company in British Columbia. The impact of the worldwide depression of the 1930s on the Canadian economy meant BC Packers was near bankruptcy as MacMillan entered that corporation. At this time of feeble economic activity in Canada, he had other worries such as the very survival of his lumber interests. In the process of financially restructuring BC Packers, however, MacMillan not only saved this Company, within a few years he gained the controlling interest, and in 1933 became its president.

The Marijean

His interests in the sea extended far beyond its natural resources and its commercial value. In 1948, he bought a 137-foot yacht and named it *Marijean* after his two daughters, Marion and Jean. His spacious yacht was comfortably furnished with a fine library, and each winter, he and his family took an extended southern vacation. MacMillan frequently invited a variety of marine biologists to accompany him on his annual voyages to southern seas – over a period of fourteen years there was one person who was invited each year (except one), Dr. Murray Newman. His first invitation came in 1954, when Newman was a marine biology post-graduate student at UBC and they continued long after he became the director of the Vancouver Aquarium in late 1955.

Newman, in his book *Life in a Fishbowl: Confessions of an Aquarium Director,* speaks appreciatively of MacMillan. MacMillan not only provided an annual opportunity for scientists to observe and research tropical sea life from his boat, but he enabled them to collect 80,000 specimens to be studied and exhibited at UBC and at the Vancouver Aquarium. MacMillan became a generous benefactor of the aquarium and was largely responsible for its construction and support. Furthermore, he introduced business leaders to the merits of the aquarium and, subsequently, some of them became supporters and directors. His own family has also taken a keen interest in the aquarium and has supported it financially as well. Newman noted that MacMillan was not a typical deep-sea sport fisherman. "He had a genuine curiosity about nature and had made a point of getting to know some famous naturalists in his day,"[9] some of whom he invited on his boat.

Over the years, many people shared in the luxury of MacMillan's voyages off the west coasts of North and South America. He enjoyed their company, their friendship, and their conversations. He chose these occasions to pursue his interest in some of the deeper issues of life. Newman observed that his "strength of personality" usually prevailed

Appointed in 1955 as director of the Vancouver Aquarium, Murray Newman travelled to Mexico and South America each year with MacMillan on his yacht, the *Marijean.*

and "wonderful conversations" would unfold aboard his boat. He challenged guests, "to face up to the moment, show spirit and accomplish something, however small, each day of their lives ...We wanted to impress him with our purposefulness. As a man of action," said Newman, "he stimulated action in those around him."[10]

Appreciation of friends

In order to show their appreciation, MacMillan's yachting guests – including the scientists, business associates, and friends – occasionally gathered for a black-tie dinner to offer their thanks. On September 29, 1960, his guests from 1948 to 1960 gathered at the UBC Faculty Club to enjoy a south-seas banquet of exotic dishes and they passed around the artfully prepared dinner program for signing. Their signatures represented a veritable who's who of British Columbia's social, educational, and business life.[11]

One guest at the banquet that night was Peter Larkin, a scientist who accompanied MacMillan on many trips. Larkin was not only a brilliant biologist; he was a witty and congenial presence who added much to the conversation. At the banquet, he offered a toast to MacMillan:

H.R.

The Dons of Spain first trod these shores
Five hundred years ago, or more,
With plumed helmets and emblazoned arms
They swaggered up the shore,
None but the birds to see and be impressed
With Cavaliers, toledo armed, well dressed.

Time passed, the clock has moved ahead,
Conquistadors have gone and in their stead,
Armed with a Polaroid, shorts, shirt and old straw hat
Amongst the birds he stands
No guns, no cutlass, just a mail of tan
Not overdressed but still a first class man.[12]

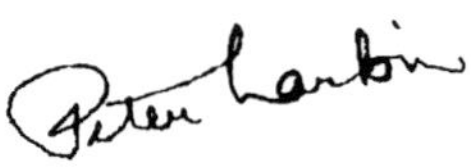

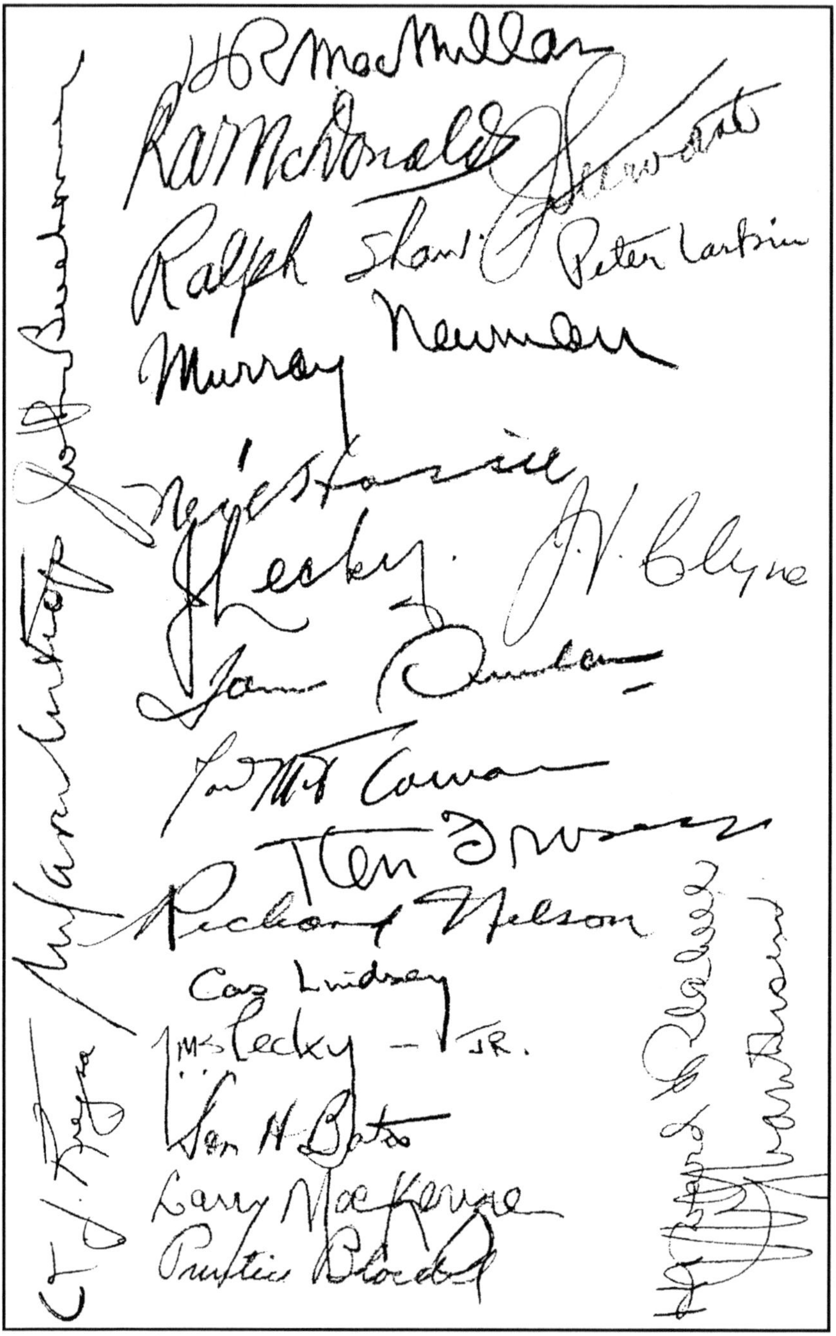

The signatures of friends who hosted and honoured MacMillan at the UBC Faculty Club on September 29, 1960.

Contacts around the world

Although MacMillan appeared to the public to be a quiet and insular man, he had, in fact, hundreds of close friends and associates around the world. He belonged to five clubs in Vancouver: the Vancouver Club, the University Club, the Terminal City Club, the Timber Club, and the Royal Vancouver Yacht Club. In addition he was a member of the Union Club in Victoria and the Canadian Club in New York City. He kept contact with many of his friends and acquaintances through his extensive Christmas card mailing. This annual seasonal greeting usually highlighted some aspect of his many interests. For example, for a number of

The cover page of the MacMillan Christmas card 1955.

years his cards featured BC and Canadian artists, thus their art was promoted around the globe.

In 1955, MacMillan's Christmas card revealed an interest not often disclosed by his peers: it reflected his deep respect for Native peoples of British Columbia by featuring a West Coast Native theme – a First Nations person standing by a canoe, with paddle in hand. Appearing in the centre of the card, the Lord's Prayer was printed in the Chinook Jargon together with an unusual English translation of this prayer.[13] Chinook Jargon was the 19th century *lingua franca* of the Pacific Northwest used for cross-cultural communication, particularly for trade negotiations between Native tribes and Europeans. It was a pidgin language based on Lower Chinook, other Native languages, French and English, and was commonly spoken in British Columbia, Oregon, Washington, and Alaska. Some BC people use remnants of the language today, e.g., *skookum* meaning *strong*.

At least one recipient of MacMillan's card in 1955, H.S. Berryman, responded to it. He was a former manager of the Victoria Lumber and Manufacturing Company of Chemainus, and the man whom MacMillan appointed, in January 1956, to be the president of MacMillan Bloedel. Berryman's response matched MacMillan's card with a note in Chinook Jargon. Berryman wrote:

> Dear H.R.,
>
> HYAS TYEE KPA KONOWAY STIX ELIP LOLO KPA CHURCH. NIKA KLAP MIKA DELATE GLOOSH CLISIMAS WAWA—KIKA WAWA NESIKA TUM TUM":
>
> *Big Chief of all our Timber before they are taken to the water (Big Chief of the Forest). I have received your very good Christmas words—it talks our opinions.*[14]

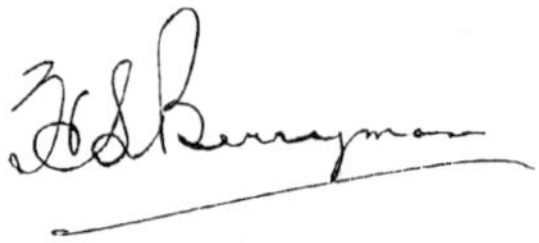

By "our opinions," Berryman may have been alluding to either their mutual understanding of the teaching of Jesus or to the English translation of the Lord's Prayer from Chinook Jargon. However, as MacMillan's faith was of a practical nature, Berryman's reference was more likely to his own appointment as president of MacMillan Bloedel and the future of the company now entrusted to him: "... send away from us all evil."

Personal sacrifice

Soon after World War II began, MacMillan's desire to respond to a great variety of requests for financial and personal support took a peculiar twist. An invitation came from the Government of Canada, which asked him to join the team assembled to lead the country's response in its declaration of war. His response was immediate. MacMillan was committed to the war effort, and he demonstrated his conviction that it was incumbent on all citizens to serve their country when it was at war. He was able to lend his organizational skills, management talents, and all of his personal time. He knew that, without his daily hands-on leadership in British Columbia, his own business interests would suffer. Despite that, he signed on as one of the "dollar-a-year" men, as they were called, and became the director of the Wartime Requirements Board – a role to which he gave his all.

MacMillan was the hard-hitting director, or president, of successive commissions or bodies the Canadian government established to deliver products needed for the war – wood, munitions, and aircraft. He did not suffer fools gladly and many feared this "red-hot poker," as some called him. Politicians either ducked or confronted his decisions, depending on their own aspirations and journalists squirmed at his often abrupt responses to their requests for clarifications about his work. "I haven't time to educate you," was MacMillan's ruthless put-down. He believed that every dime he had worked so hard to gain for the war effort would be lost unless the war was won. "A man like that, who has all the ability MacMillan has, should be kept even if it is hard to bunk with a wildcat. There is a time, after all, when ruthlessness has its very excellent virtues."[15] Some suggested that he be appointed to the cabinet, which would give him the power needed to execute his vision of how Canada should respond to the war effort. Some thought that MacMillan was the heir apparent to the Conservative Party leadership, while others found him to be too dangerous and therefore should not be given too much power. The truth was that both Liberal and Conservative speculators were left without any indication of what political party he would support, despite the fact that he had once written a reform speech for R.B. Bennett, the Conservative Prime Minister.

Canadian buzz saw

An example of MacMillan's matter-of-fact approach to religion was cited in the April 1, 1941, issue of *Time* magazine. Under the headline "Canadian Buzz Saw," the article reported on a formal dinner where there had been a hushed lull while the diners waited for someone to say grace. MacMillan broke the silence with his booming voice, "MacMillan

Hard-driving, hard-headed MacMillan was WW II director of merchant ship building and shipping of the Government of Canada in 1942.

speaking!" Clearly, he felt comfortable with the role of offering a public prayer of thanksgiving and blessing.

The same article in *Time* included another reference to MacMillan and the Lord's Prayer. When *Time* magazine interviewed him, he was the wartime director of merchant shipbuilding and shipping of the government of Canada. The article noted that he was like a "buzz saw" as he cut through red tape, a "hard-driving, hardheaded lumberman who believes in getting things done. No business-as-usual fuddy-duddy." His manner was not always smooth, however, and to a muddle-headed government clerk who telephoned him to ask what should be done with a carload of shingles, he replied, "Print the Lord's Prayer on every one of them!"

We have what it takes

MacMillan was not only in the thick of leadership and decision-making in Ottawa's war effort, he was on a campaign trail to champion the sale of Victory Bonds. He delivered a powerful speech to the people of British Columbia on May 30, 1941 and Premier Thomas "Duff" Pattullo was in attendance. MacMillan had called everyone there to show unfailing loyalty to, and financial support for, the "largest loan ever undertaken in British Columbia." As he thundered his message, he assured his audience, "WE HAVE WHAT IT TAKES ... " And what did it take? "Organizing ability, intelligence, faith, wealth, and character," he said.[16]

His speech revealed the depth of his faith and his understanding of the social order in which he was nourished and raised. "There are certain principles which are essential to life as we wish to live [it] and see it lived," he said. "Those principles are now in a dire period, and we believe, therefore, that we are truly fighting for our lives, since life to us is worthless if the principles on which it is built are to be destroyed. First, the religious principle of the absolute value of every human soul. Second, the moral principle of respect for personality and for conscience. Third, the social principle of individual liberty."[17]

As he elaborated each of his religious, moral, and social principles in his forty-five minute Victory Bond speech, MacMillan adopted the stance typical of a zealous evangelical preacher. He raised his voice, waved his fist, pointed his finger, pounded the lectern, and used all the techniques common to a raving evangelist. MacMillan's compelling and emotional appeal was aimed at persuading his audience to buy victory bonds, which addressed the need for greater industrial production in the war effort. Called upon to make a sacrifice akin to those on the front lines of war, his listeners had a clear obligation. MacMillan laced his speech with quotations from the Bible. "The Bible tells you that faith is the substance of things hoped for, the evidence of things not seen," he thundered. "Do you agree that you are imbued with faith as are those others of our race now on the blazing firing line? Their faith is in us – that we will not let them down." Then he hammered his point home: "The Bible also says that faith without works is dead. This loan is your today's opportunity to prove your faith by works."[18] By the end of MacMillan's speech, everyone in the audience was aware of the need for their personal and immediate sacrifice.

Winning ways

There was another side to MacMillan, a smoother quality, a trait that won hearts and commitments for his business enterprises. His charm, engaging personality, and intuitive sense of opportunity attracted decision makers in the business world. Together with VanDusen's export trade acumen, the pair proved unbeatable. "Nothing is impossible," MacMillan would constantly say. Their combined tactic of acquiring smaller lumber enterprises included, predictably, the Victoria Lumber and Manufacturing Company (VLMC) in 1946. Despite the best efforts of its owner, Tom Humbird, to prevent the sale to MacMillan, he was outwitted. MacMillan secretly engaged his friend, E.P. Taylor, to make the purchase and sell it to MacMillan. Humbird was furious! There must have been a measure of personal delight and satisfaction for MacMillan when he signed a cheque for nine million dollars to Taylor and again, as he walked through the VLMC office door for the first time since 1917. This time, however, he was its owner.

J.V. Clyne relayed his memories remembrances of this incident in *The Vancouver Sun* on October 30, 1964, "Unfortunately," he said, "several days before he [MacMillan] returned to visit the mill, the door had been removed, having become old and worn out through wear and tear – probably having been slammed too often by disgruntled employees. With a true sense of historical accuracy, H.R. ordered the door to be found and replaced, which was done, and as far as I am concerned it will remain there until it falls off its hinges." This original door remains in place to this day and its exact copy, installed next to it, is used as the main entrance to the office. Following MacMillan's takeover of the mill, the word "Manufacturing" vanished from its name and it became simply known as "The Victoria Lumber Company."

The "team," as VanDusen and MacMillan called themselves, outsmarted their business rivals. Their insatiable energy and drive combined to make them, and MacMillan especially, among the most widely known businessmen in the country. By the end of his business and professional career, he was the major shareholder of MacMillan Bloedel. *Time* magazine declared him the "world's No. 1 exporter of lumber."[19] He had become one of British Columbia's most powerful individuals, a person of vast wealth and influence.

A third eye

There is no question that MacMillan's achievements as a professional forester and business leader were outstanding. By any standard, few people can match his achievements. One of his special gifts was an uncanny talent to predict a future event and to act on what he felt he could foresee. Once he had assessed an occasion and its implications, nothing seemed to be impossible for him and his assessments included far more than the intimate knowledge of the business details. Through his reading, personal observations, and global contacts, he had a thorough understanding of world events. In short, he had an appreciation of where and when trade, economic conditions, and political realities indicated that a business opportunity was opening in which he should invest his energies and capital resources. The timing of an investment was crucial and, when all indicators suggested that it was time to make a business move, he acted with speed – hesitation would surely result in loss.

MacMillan could analyze people with a perceptiveness that often eluded others and he thought through the implications of a decision before making it. He knew how to conceive a plan and execute it. Long before Tom Peters' *In Search of Excellence* became a popular text and business guide during the 1980s and 1990s, MacMillan exemplified the effective use of business ideals and management principles he articulated. MacMillan was guided by his own advice to others: "Analyze, organize, deputize, supervise, energize and, if necessary, excise ... Failure in any of these steps," he cautioned, "weakens or ruins the enterprise. If anyone fails in these essential practices, correct quickly with [the] best possible judgment."[20]

MacMillan understood his business was a living entity. He cautioned his successors to "protect it from infection by those ills that weaken so many businesses."[21] He went on to illustrate his point by citing the many logging and sawmill companies that owned mills or timber on the BC coast and had flourished when the H.R. MacMillan Export Company started but that not one survived with the same ownership. These failed companies had fallen victim to creditors or estates and had been sold to newcomers or to stronger competitors.

MacMillan knew, however, that he, too, was mortal and that it was necessary to plan the disposition of his accumulated resources to meet his own needs, to care for his family, and to give back to the community. The time had come to share some of his vast wealth. Perhaps in doing so, he would set the pace and offer an example for others to follow while fulfilling his own philanthropic mission.

Chapter Six

A Time to Share

There never seemed to be a time in MacMillan's life when he was not in the lead. He set a challenging pace for himself and for those he associated with. Whether we observe him as a brilliant student, as a dedicated career forester, or as a successful businessman, he was a leader and an exceptional one at that.

In his professional career, he was one of the first foresters to articulate a comprehensive stewardship of the natural timber resources of North America. If the harvest was to continue, he argued, it was necessary to implement a conservation ethic. In his early student days, and as editor of *The OAC Review,* he espoused a scientific approach to the commercial harvesting of forests. Later, in his business career, he led by example, not only in his responsibilities toward the forests, but in how to care for the safety and welfare of his employees. He supported the communities where his company was located and provided amenities for their recreation and social needs.

Natural-born leader

Although MacMillan was personally accountable for his own life, he also felt a keen responsibility towards the community that had given him so much. That sense of responsibility came early in life, at a time when many young people think primarily of themselves and the pursuit of their personal interests. In November 1949, *Canadian Business* magazine published an article entitled "The MacMillan Story," which drew attention to his leadership and ability to organize, even as a student. The author, A. Harris Mitchell, describes how MacMillan, working in a summer forestry camp, headed a committee that supported a small

Chalmers United Church on West 12th Avenue, Vancouver, BC.
Credit: Jost Tieller, Artist.

community church. One Sunday, the travelling missionary was unavailable so MacMillan organized the service, conducted it, and served as preacher. He was twenty-two years old at the time, and perhaps this pulpit experience was the beginning of his public-speaking career.

Two-phased philanthropic career

On February 10, 1965, *The Vancouver Sun* announced several major gifts MacMillan had made to the University of British Columbia. In an article describing these donations, staff reporter Moira Farrow suggested that MacMillan had began this "new career" ten years earlier, in 1955. What the *Sun* was really reporting, however, was a second phase of his philanthropic career, the first phase of which had been underway for at least forty years. Both phases of his philanthropy are noteworthy for MacMillan's close involvement in managing the whole of the gift-granting process: he carefully studied each request, he asked questions for

Harvey Reginald and Edna MacMillan were communicant members of Chalmers United Church, Vancouver.

clarification, he took an interest in the proposed recipient of his benefaction, and he made the final decision on granting an award. Finally, he signed the letter accompanying the financial gift and wished the recipients well with their work. By 1965, MacMillan had already established a considerable record of providing financial support to small organizations and large gifts to major institutions.

First gifts

For four decades his offerings to Chalmers United Church in Vancouver were the largest of any member. This was the same church where his mother, Joanna had devoted much of her life to the congregation's missions and outreach programs. In the 1920s and 1930s, MacMillan's Sunday offering was $500 each week, a very large sum at that time.[1]

Beyond this commitment, his first philanthropic gift to an educational institution was in 1927. He furnished the library at the newly founded Union College, which was the training facility for seminarians of the United Church of Canada that was located on the grounds of the University of British Columbia. In the case of furnishing the Union College library, he was totally involved: he corresponded with the architectural firm, G.L. Thornton Sharp & Charles J. Thompson, regarding the design and construction of tables and settees; he selected the edge grain lumber for kiln drying; and he ordered the materials for delivery to the Restmore Manufacturing Company for the cabinetry work.

The gift of furnishings to the theological college library is MacMillan's first indication of where his philanthropic priorities would lie. It is not surprising that he supported libraries, as he was a scholar and reading was a priority for him. For MacMillan, higher education was a core value, and during his philanthropic career he directed the majority of his donations to educational institutions. Opportunities for post-secondary education in the arts, sciences, and theology were important to him. He felt that the subsequent leadership provided by their graduates enriched democratic society. In this belief, he anticipated what, by the end of the 20th century, would become an immensely complex moral and ethical relationship between science and theology.

The first award MacMillan made to the University of British Columbia was in 1929[2] – it was a gift of $1,000 for a scholarship in economics. He described it as a symbol of his gratitude for the profits he had made in his first business enterprise, the H. R. MacMillan Export Company. The award included a free study trip to Asia on his shipping line.

MacMillan's gifts from 1927 to 1929 were far more than an expression of gratitude, they were an early sign of his lifelong commitment to philanthropy – a commitment that would faithfully unfold in the succeeding years. That personal covenant, an element of his rule of life, remained with him to the end of his days.

Needs to be met

For many years, MacMillan and his close friend VanDusen discussed possible funding options to meet the needs of a variety of institutions. Together, they influenced other potential donors and, of course, had an impact on the social, religious, and educational organizations that were the recipients of their largesse.

Both MacMillan and VanDusen worked together on the initial development of the Vancouver Welfare Federation. In 1931, just as the national economic depression was becoming worse, MacMillan contributed $150 to the Federation. His gifts increased to $2,000 and $3,000 each year during the time of Canada's greatest economic need.

In MacMillan's first phase of financial gift-making, he became widely known as a potential source for funding small social and humanitarian projects. Some people in need of funds sent a letter of appeal to "the great milk cow," as one institutional head insensitively described MacMillan. But he was able to see through comments of this kind with ease, and if he detected even the slightest hint of presumption in a request, he would either refuse it or respond with a minimal gift. Despite the occasional refusal, he was well known for his generous responses to the many and varied requests that crossed his desk, each of which he handled himself.

MacMillan's compassion extended to even the heart-felt plea for a very small gift. Perhaps the two smallest "grants" he ever made were for two dollars each. In 1940, a request came from an artist in Cache Creek who asked for help to pay for paints and equipment. Another letter came to him from a man in the BC interior who needed a pair of boots so that he could work. MacMillan recognized their motivation, and could hardly wait to send a cheque in either case.

Founding a foundation

In 1944, an event unfolded that would ultimately shape much of the future investment in religious, social, and educational institutions in BC. Drushka records the dramatic founding of the Vancouver Foundation when:

> ... a shy and reserved woman with no close relatives, Alice MacKay, gave the federation* $1,000, with the stipulation that only the interest it earned be used each year. At about the same time, VanDusen had organized a body called the Vancouver Foundation to support the Welfare Federation's operating expenses. He conceived the idea of using Alice MacKay's $1,000 to start an endowment fund to support various charitable activities. Realizing the interest on her donation would accomplish little, he went out and persuaded nine others, including MacMillan, to join him in mak-

---

* Vancouver Welfare Federation

VANCOUVER FOUNDATION

Logo of the Vancouver Foundation that MacMillan helped to found with a gift of $10,000 in 1944.

> ing a $10,000 donation each to the fund. This was the beginning of the Vancouver Foundation, which within fifty years became Canada's largest community foundation, with endowments of more than $350 million and annual grants of more than $22 million. Although MacMillan was an early and major supporter of the Vancouver Foundation, it was really VanDusen's creation.[3]

Today, the Vancouver Foundation is the largest community foundation in Canada.

A sacred natural resource

Not long after World War II in 1946, just after MacMillan had acquired the Victoria Lumber and Manufacturing Company, he entered into correspondence with The Honourable John Hart, the premier of the province. The matter under discussion was the possible transfer of one of the assets of the VLMC to the province – what is now commonly known as Cathedral Grove near Cameron Lake, en route to Port Alberni on Vancouver Island.[4] Initially, there were negotiations regarding the terms of exchange, but MacMillan finally, agreed to give the entire 332-acre tract of original old-growth forest to the province to be cared for in perpetuity for the benefit of all. The only stipulated conditions regarded taxes on the land and the establishment of boundaries. Otherwise MacMillan unconditionally gave the land to the people of the province and Premier John Hart graciously accepted the gift on behalf of all British Columbians. The provincial media affirmed the government decision that the magnificent stand of old-growth Douglas fir be named MacMillan Park.

A birthday gift to BC

It was not until 1955 that MacMillan made personal use of the Vancouver Foundation to establish the H.R. MacMillan Education Fund and it gave him another opportunity to be personally involved in the transfer of resources to worthy community projects. This major gift of $500,00 to the Vancouver Foundation[5] marked his 70th birthday. The trust was to be disbursed in no less than twenty years and no more than thirty years. MacMillan intended the money to benefit departments of UBC dealing with forestry, fish, ecology, and conservation. The newspapers mentioned the birthday gift in the following headline:

> **"$500,000 OPENED CAREER OF GIVING**
> **MacMillan's Birthday marked by 1955 Gift to Foundation."**
> (*Vancouver Sun,* February 10, 1955)

To mark his seventy-first birthday, he gave a further $50,000 toward advancing studies of the economic realities of provincial industries and resources.

Support for the community

MacMillan's philanthropic donations grew in number, amount, and frequency. Sometimes an announcement of a gift or endowment came with little notice – his own family often learned of his generosity for the first time by picking up a copy of the morning newspaper. The press in the 1950s and 1960s were eager to feature MacMillan's philanthropy as he made yet another major gift. For example, headlines read:

> **"MACMILLAN GIVES B.C. $2 MILLION-"**
> **Trust Fund to Aid Education, Welfare"**
> (*The Vancouver Sun,* August 8, 1962)
>
> **"3rd MILLION GIVEN TO CHARITY**
> **by B.C. Lumber Magnate"**
> (*The Vancouver Sun,)* Dec. 5, 1962)

"H.R. MACMILLAN DONATES $8.3 MILLION to University of British Columbia"

(*The Province*, Feb. 10, 1965)

A rule of life

Following the merger of the BC forest-product giants H.R. MacMillan Export Company and Bloedel, Stewart and Welch in 1951, MacMillan sold shares in the newly formed company, MacMillan Bloedel Limited, to support his charitable pursuits. During this period, he took a disciplined approach to financial gift giving – or a rule of life, as the Anglican *Book of Common Prayer* suggests. He committed himself to give at least ten percent of his annual net income to welfare, education, and religious enterprises. But beyond that amount, he made further personal donations and gave gifts from his corporate funds.[6] For example, at the time of UBC's 1957 campaign for funds, MacMillan Bloedel donated $150,000.

In 1960, he established The H.R. MacMillan Family Fund within the Vancouver Foundation. This gave him and his family, as donors, the right to advise on the allocation of the fund for educational and welfare purposes in British Columbia.

A visceral attachment

MacMillan gave little outside the province of BC, and his responses to requests from the universities he had attended (Ontario Agriculture College, The University of Toronto, and Yale University), were rare and small by comparison to those he made to the University of British Columbia. He had a fierce loyalty to BC, a "visceral attachment," as one author described it.[7] He felt that the older and more-established universities in eastern Canada and the US had a wide and prosperous support base, but that the younger BC institutions needed his help. He gave to churches and causes in the communities where MacMillan Bloedel had forestry operations. His love for this province extended to his desire for its people to have access to good libraries. Almost every small coastal town or village in British Columbia can claim that its local library had benefited from MacMillan's largesse. Libraries were a perfect fit for his philanthropic goals and personal interests.

Handwritten notes

MacMillan was a hands-on person – he read every letter of appeal for funding. In response, he noted his decisions or instructions on each letter for his long-time personal secretary, Miss Dorothy E. Dee. Occasionally, he penciled in instructions for her to request further clarification or information. Most often than not, he simply noted the gist of his disposition toward the appeal, and he always added his initials, H.R.M. He would carefully note the salutation Miss Dee was to use, and she would then compose and type the letter of response. There was no need to complete a long application form for donations or grants from MacMillan. Nor were there bureaucratic hoops to jump through or references required; indeed, there were none of the many obstacles today's applicants must face when applying for grants from foundations and trusts. Rather, a simple letter was usually sufficient in drawing his immediate and personal attention.

Nevertheless, MacMillan was a stickler for detail and correct form – a quality formed by his love of reading and writing. He expected a standard consistent with what he perceived to be the author's education and status in the community. An example of this expectation appears in the first words of his letter to the rector of a Victoria parish, which announced his refusal of a grant: "I have your undated letter ..."[8] as if to say, "Before you mail your letter, be sure to check it for proper form!"

Dorothy E. Dee, the highly competent and long-time secretary to MacMillan. In appreciation for her services, he left her a life-time monthly income.

Keep it personal

He insisted that letters he received be personal in nature. He would not have taken kindly to 21st century appeals sent to millions of Canadian homes on behalf of charitable organizations. This impersonal style of the form letter did not appeal to MacMillan.

In 1969, a form letter from the University of British Columbia reached his desk, which made him feel his generosity was being taken for granted. It was a letter from the University's Wills and Bequests Committee suggesting he might want to include the university in his will. His immediate and stinging response made its point, "I do not know what I shall do, therefore I suggest to your committee that they contain themselves in patience until, if ever, they find out."[9] UBC did discover the truth in due course and it was a difficult one. To his beloved alma mater he left no final financial gift, endowment, shares, or money. He did leave, however, his collection of rare books to UBC to be designated the "H.R. MacMillan Rare Book Collection."

Big saw cuts

Attempts at humour were not necessarily helpful to applicants either – MacMillan was not amused by witless commentary. In 1951, the following letter arrived from Alberta asking for financial assistance:

> Dear Mr. MacMillan,
>
> Wondered if you would be interested in helping an old lumberjack get hold of a few good beef steaks?
>
> If for this letter you do not thaw,
> Just run it through the big, big saw.[10]

Figuratively speaking, he did just that and the applicant's letter was shredded. In his courteous and sensitive letter of rejection he wrote, "It is best for me that I give help to the cases in our own territory where I have some knowledge of what is going on and can see the results."[11]

An expanding philanthropy

In the early 1960s, the second phase of MacMillan's philanthropic career continued to expand. His increasing generosity became more widely known in the community and charities and educational institutions took notice and approached him for financial support. He rarely dealt directly with official development officers. Rather, he met with those with direct responsibility for the institutions or its major departments in need.

MacMillan's passion for giving stemmed from his sense of responsibility for the community: "I consider it proper that those who have benefited by such resources as forestry and fishing should give evidence of their interest in perpetuating and advancing the welfare of those industries."[12] It was a time to lead by example, as he had always done. He became one of the chief supporters of the Institute of Fisheries at the University of British Columbia, and in 1965 he gave the Institute $750,000, clearly stating his intentions for its use:

> This money would be used primarily for attracting and keeping the best teaching staff in the Institute of Fisheries ... [and] that any portion of the money used for buying equipment would be kept at a minimum – my thought being that our governments, which spend millions on equipment in biological stations and research vessels, should provide for the University such equipment as is necessary to train the scientist.[13]

It should be no surprise to discover that he gave liberally to forestry concerns. For many years, he provided for the salary of the University of British Columbia's first dean of forestry, and he established a variety of scholarships and programs in forestry. Although he also expected other benefactors to assist in the maintenance of these programs, they did not follow his lead immediately. When he discovered that his donations accounted for fully one-fifth of the school's total budget, he began turning down further requests. However, in 1965, he gave three million dollars to assist in the building of the forestry-agriculture building at UBC. When it opened on June 14, 1967, it bore his name, the new H.R. MacMillan Building.

The financial gifts mentioned above represent only a limited account of MacMillan's gifts. Furthermore, the variety of causes he supported was extensive. For example, he gave $50,000 to the Alberni Valley and West Coast Disaster Fund for victims of the tidal waves following an earthquake; $10,000 for scholarships "to school children where the business operates"; $25,000 to the Vancouver Aquarium for a cold saltwater supply for their fishtanks; and from 1959-64, 160 separate projects in several UBC faculties.[14] "I regard the UBC," said MacMillan, "as an institution which has a greater potential than any other for making a contribution to the future of the population of this province."[15] The foregoing list of projects illustrates that, in addition to the multi-million dollar endowments, he also gave many relatively small gifts. In his view, the long-term impact of the small gift also represents a sound investment. Even so, MacMillan was about to change his philanthropic strategy – a second and dramatic phase of sharing his wealth with the people of British Columbia was unfolding.

Chapter Seven

A Time to Retire

By the late 1950s and early 1960s, the circumstances of MacMillan's life had changed dramatically. In 1958, MacMillan had appointed J.V. Clyne as the Chair of MacMillan Bloedel. In some ways, it was a curious appointment – choosing someone with no business experience to effectively lead a very large company. Clyne had neither MacMillan's experience nor his expertise, and he did not lend his unconditional support to the business principles that had proven to be so effective for MacMillan. As a result, he instituted new directions for the company to replace those previously articulated and implemented by MacMillan. Furthermore, communication between the company's newly appointed directors and the old guard was not as frequent or informative as MacMillan would have wished. He wanted and expected to be included in the making of decisions, yet up until the late 1950s, his relations with the company became increasingly strained. His interventions were not all received positively by the Board of Directors and he came to feel distanced from the company he had founded, shaped, and directed over many decades.

For many people, retirement brings with it a sense of powerlessness – be it physical, intellectual, spiritual, political or economic. For MacMillan, this new and deeply emotional reality was an unsettling one that he found difficult to accept. Retirement sharpened his leadership sensibilities and his identification with the operation was unraveling rapidly. Closely tied to the fortunes of his company over many decades, MacMillan's loss of involvement and influence were deeply troubling to him, and his inner spirit grew demoralized. As his energies began to flag, he withdrew from the centre stage he had once occupied in the forest industry. By 1961, it was becoming apparent that none of his grandchildren wished to carry on his business interests and his daughters Marion and Jean, with their growing families, had demands of their own. In any case, he had always declared it to be a fundamental principle of his business practice that there be no nepotism.[1] At the age of seventy-six, he decided it was time to retire.

Family matters

Not least among MacMillan's concerns was Edna's declining health. As she was being cared for at home, her rapid deterioration was a daily reminder to him of the shortness, fragility, and uncertainty of human

Edna Mulloy MacMillan, not seen often in public, was homemaker and mother to two daughters. She joined MacMillan on his yacht for trips to Mexico and for informal gatherings with close friends.

life. It was not easy for MacMillan to see his beloved partner deteriorating. Edna, his wife of fifty-one years, died January 20, 1962.

Things of the spirit

It was in this period of great personal change, upheaval, and despair that MacMillan became even more interested in eternal questions. He read poetry and he read dozens of books. He cut and kept clippings of distilled wisdom and spiritual insight from his favorite authors and philosophers. Gradually, his distress over the new directions of "my outfit," as he called MacMillan Bloedel, began to diminish. Perhaps it was then that he meditated on the insights of William Blake, the late 17th and early 18th century artist, poet, and visionary, a quotation of whose remained treasured to MacMillan until the time of his death:[2]

> I was angry with my friend: I told my wrath, my wrath did end.
> I was angry with my foe: I told it not, my wrath did grow.

He had his secretary type particular quotations he treasured, which seemed to provide him with the understanding or comfort he was seeking. For example, this well-known Persian proverb remained near his desk:

> He that knows not and knows not that he knows not is a fool, shun him,
> He that knows not and knows that he knows not is a child, teach him,
> He that knows and knows not that he knows, is asleep, wake him,
> He that knows and knows that he knows, is a wise man, follow him.[3]

Christian baptism

MacMillan's search for answers to eternal questions is what led him to seek out the details of his baptism, but he found no record. This disturbed him, and he took his concerns to The Reverend Robert (Bob) Damasus Payne, the son of an Oregon business friend. Payne was a member of the monastic community of the Benedictine Abbey in Mission, BC. For some fifteen years, between 1960 and 1976, he often visited MacMillan in his home at 3741 Hudson Street, in the Shaughnessy district of Vancouver. Payne, a keen photographer, avid hiker, and risky mountaineer, wrote frequent and lengthy letters to MacMillan about his adventures in the mountain valleys on the north and south sides of the Fraser River. During their visits, MacMillan and Payne shared lively conversations on topics of mutual interest. Occasionally, Payne overstayed his welcome, and the family would dispatch their dog into the den or living room when they thought it time for him to conclude the visit. The fact that the dog was carrying Payne's hat was signal enough that the good priest should leave.

Payne, whose father was baptized by a nurse shortly before he died, seems to have heard about MacMillan's interest in baptism through Sister St. John, a member of the Order of the Precious Blood, whose convent was located close to the MacMillan home. The sister had prayed for a long time that MacMillan would receive the sacrament of baptism. During one visit, the conversation between Payne and MacMillan turned to his need for baptism, and Payne entered into what he later described as a short course on the meaning of the sacrament.

The course offered by Payne may have included a reference to the philosophy, theology, and life experience that led St. Augustine of Hippo (354-430 AD) to baptism in 387 AD at the age of thirty-two. He was a professor of rhetoric at Tagaste, Carthage, and Rome before his conversion to Christianity. Like MacMillan, Augustine was highly influenced by his mother.

When Payne realized that his elderly friend MacMillan possessed the required faith and repentance, he proceeded to administer the sacrament. He described the baptismal event:

The Reverend Fr. Robert Damasus Payne on Black Tusk Mountain near Mission, BC, June 1974, shortly before he died in a mountain fall while shooting photographs. He baptized MacMillan in the early 1970s.

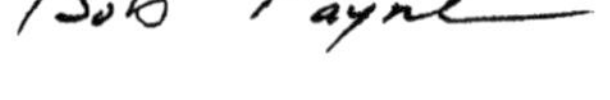

> So I baptized him without ceremony one day in his front room with some water in a fishbowl. I didn't ask him if he wanted to be a Catholic, but if he wanted to be baptized. It was a short course, like the one Philip the deacon gave to the Ethiopian on the way down from Jerusalem.[4]

On the day following MacMillan's death in 1976, Payne celebrated a mass in the Chapel of the Poor Clare Sisters in Mission, BC, in thanksgiving for MacMillan's life. In his homily, Payne shared the story of MacMillan's baptism. He spoke of Augustine and about the influence of St. Monica, his mother. Most certainly, the instruction Payne provided MacMillan leading up to his baptism included a reference to the first-century meeting of Philip with the Ethiopian. In this story, recorded in *The Acts of the Apostles* (8:26-40), Philip discerns the eunuch's faith to be present and, as water was nearby and all elements required for baptism were present, he baptized him immediately.

Considering MacMillan's great love for the sea and his no-frills approach to matters of faith, it seems appropriate that he was baptized with little ceremony and with water "in" a fish bowl.

Ecumenist

MacMillan had a high respect for the many expressions of Christian faith and tradition that had influenced his own life. As a child, several denominations formed and shaped his spiritual life, but the disciplined influence of the Society of Friends was a primary one. As a young university student, he had made his personal testimony of faith and had been confirmed in the Presbyterian Church. Apparently, the question as to whether he was baptized or not, was not raised at this time. As an adult, he and his wife were communicant members of the United Church. Then, toward the end of his life, a Roman Catholic priest baptized him. In his latter years, he and his family were associated with the Anglican Church, and it was from the Anglican Cathedral that he was buried some years later. In his personal journey of faith and through his generous support of different denominations, MacMillan recognized the essential unity of the church, despite its variations.

More than anything else, MacMillan's financial gifts to a variety of denominations illustrates his all-embracing approach to the Christian faith as a whole and the range of gifts given to churches serves to illustrate this point. In 1959, he supported St. James' Anglican Church in Vancouver[5] with its Christmas Cheer Fund, and the following year he

In retirement, MacMillan sits relaxed I his home library at 3741 Hudson Street, Vancouver, BC.

provided the Roman Catholic Benedictine Monastery in Mission with a grant of $5,000 toward the building of its new gymnasium. In 1964, he gave $1,200 toward the restoration of St. Anne's Anglican Church in French Creek on Vancouver Island, near his Qualicum Beach summer home. He gave to the Shantymen's Christian Association, an evangelical group working in remote areas. He did not forget his formative roots in the Society of Friends or his ancestor's family church in Ontario and he responded generously to the appeal for cedar shakes for the roof of the Yonge Street Quaker Meeting House, near Newmarket. In the late 1960's, he was a regular supporter of Anglican Camp Artaban on Gambier Island, the United Church Naramata Conference Centre in the Okanagan, and the Missions to Seamen in Vancouver.[6]

Although MacMillan had long ago given up his direct association with the Quakers, it was nonetheless this branch of the Christian church that shaped so much of his openness to receive requests, and the generosity to support programs to address human need.

Greyshakes, the well-used summer home of the MacMillans at Qualicum Beach, BC.
Credit: Alistair M. Ball, Artist

The Yonge Street Meeting House of the Society of Friends near Aurora, Ontario, has been in continuous use since 1910. The cedar roofing shakes were a gift from MacMillan.

Two small books

MacMillan was a professional forester, a scientist and a hardheaded business man. What some might consider to be an unusual interest for him, namely theological education, became an absorbing passion for him later in life. Perhaps this dimension of his philanthropy began with his gift of two rare books to Union College at the University of British Columbia. One book was a volume of the sermons of Hugh Latimer (1485–1555), Bishop of Worcester, published in 1550, which addressed social injustices and ecclesiastical abuses of the day as well as major doctrinal issues. The other volume was a 1597 edition of the Holy Bible. It gave him great pleasure to give these small volumes – particularly because they were rare editions. Later, MacMillan gave a Bible printed in 1632 to the library of Union College.

Anxious about the bindings of the two rare books he had given, MacMillan corresponded with the principal of Union College, The Reverend Dr. William Stephens Taylor, to suggest the repair or replacement of their covers. On September 4, 1959, Taylor replied to offer the option of paying $30 or $18.25 to cover the costs of binding the books, depending on the quality of repair MacMillan thought appropriate. MacMillan initialed his endorsement on September 15th, "$30xx OK" and instructed his long-time personal secretary, Dorothy Dee, to address the letter, "Dear Dr. Taylor."

A passion for books

Collecting books, particularly rare books, was MacMillan's great passion. He was a voracious reader, and reading was a habit he had picked up while in the Quebec sanatorium for two-and-a-half years. He was particularly interested in old texts and manuscripts, which prompted him to collect rare theological books.

His passion for books led to an extensive correspondence with Dr. Taylor. Although Taylor had met MacMillan in 1950, their new association was made possible through the close relationship MacMillan enjoyed with W.J. VanDusen.[7] There is little doubt that the twelve- member round table of major leaders in business, education, and religion, who met frequently at the Vancouver Club for lunches and dinners, provided a fruitful forum to nourish not only significant business relationships, but philanthropic partnerships.

Whitfield Julian VanDusen, the highly competent comptroller of MacMillan Bloedel Limited, was a close business colleague and friend of MacMillan for fifty years.

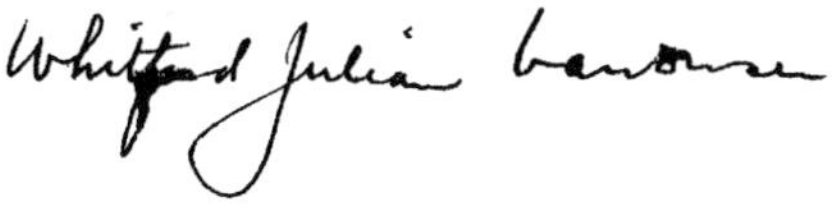

VanDusen met Taylor in 1951 while Taylor was touring the province on behalf of Union College during a fundraising drive.[8] Their relationship culminated in 1958 when VanDusen was elected chair of the Board of Governors of Union College, a post he continued to hold until 1964. VanDusen expressed his deep commitment to the College in 1961 when he personally funded the building of its new east wing.[9]

Later, Taylor reflected on VanDusen's influence in encouraging MacMillan's support of United College. MacMillan and VanDusen had many opportunities to discuss the matter during their leisurely associations and in the course of business meetings. Taylor's own description of a conversation between VanDusen and a person Taylor identified as "a prospective contributor" leaves little doubt that the latter was MacMillan:

> I suppose he thought my presence as principal might give a certain authenticity to the appeal. But it was his own status among his contemporaries that made the appeal effective. I recall one incident particularly. The interview was with a very well-known business man. Mr. VanDusen spoke of the college and its needs, directly and briefly, without bothering about any preliminary softening up process.
>
> "You are supporting this?" asked the other man.
>
> "Yes," replied Mr. VanDusen.
>
> "You are giving to it?" the other man asked.
>
> Again, Mr. VanDusen agreed.
>
> There was a pause for a few moments.

Then, "How much?" asked the other man. I thought he was asking how much Mr. VanDusen was giving. But no. There was some silent unspoken understanding between them. The other man was really asking how much Mr. VanDusen thought he should give.

"Ten," said Mr. VanDusen. I saw the other man give a slight start, and there was a momentary narrowing of the eyes. Then he was normal again.

"Done," he said.

I, rather naively, thought it meant ten dollars, or perhaps ten hundreds. I considered it not much to ask of a leading businessman. But when the cheque came in, it was for ten thousand dollars.[10]

It was during VanDusen's tenure as chair of the Board of Governors of Union College that MacMillan gave $250,000 in January 1964, to endow two faculty appointments, one at each of the two theological colleges at UBC. The first was for The Edna MacMillan Chair in Church History at the Anglican Theological College. The second was to fund the continuing Joanna Willson MacMillan Chair in Systematic Theology at Union College that he had established fourteen years earlier in 1950. At that time, MacMillan committed the H.R. MacMillan Family Fund to provide for a full-time appointment to the faculty of Union College to be known as the Joanna Willson MacMillan Professorship. His annual gift of $4,000 was to be paid for an initial period of ten years but, at the conclusion of the decade, he doubled this gift to $8,000 annually. These two professorial chairs were financed through the H.R. MacMillan Education Fund. His intention was to support these chairs for a limited time only, and to make the final payment in 1979. That year, the H.R. MacMillan Estate Charitable Trust replaced this annual gift with a new annual commitment for $22,500 to a combined chair by the name of the Edna and Joanna MacMillan Professorial Chair.[11]

In addition to VanDusen and Taylor, there was a third significant person in these ongoing friendships and conversations with MacMillan. It was Dr. Norman A.M. "Larry" MacKenzie, president of the University of British Columbia. During his student days, MacKenzie had been an active member of the Student Christian Movement and had become a close friend and collaborator with MacMillan at UBC. They often took fishing trips on the *Marijean,* during which MacMillan enjoyed poetry read to him by MacKenzie.[12]

While serving as president of the University of British Columbia (1944-62), MacKenzie was a member of the University Hill United Church that then worshipped in the Union College chapel. He was

Long-time business associate and chancellor of the University of British Columbia (1966–1969), John Buchanan, encouraged MacMillan to finance the improved library facilities of Union College at UBC.

interested in establishing a substantial theological library at UBC but, because there was no Department of Religious Studies on campus, he sought other ways to achieve his dream. Taylor later felt that both MacKenzie and MacMillan agreed to encourage the endowment of the library of Union College to benefit UBC as a whole, but it was VanDusen who encouraged them in this regard. Furthermore, VanDusen's support may have encouraged MacMillan's first major gift to Union College library in the 1950s.

At least one other person, John M. Buchanan of BC Packers, was involved in discussions with MacMillan and Taylor and all identified the need for the library resource in July 1958. In a note to Buchanan on August 6th, MacMillan signed his approval of a grant for $50,000[13] to assist the Union College library to upgrade its standards and receive full accreditation by the Association of American Theological Schools.

A growing correspondence

In response to Taylor's letters of appreciation to his benefactor, MacMillan's notes were brief. He simply acknowledged the letter from the principal and added the following, or similar words, "I am, of course, very pleased that you will find my gift useful. With kindest regards, and all Good Wishes for the Year we have just entered. Yours sincerely, H.R. MacMillan."[14] Unknown to both men, a pastoral ministry was unfolding in their relationship through the correspondence. Taylor later reflected that at no time had he any sense it would lead to an unexpected, and highly significant gift to the College.

MacMillan and Taylor corresponded at intervals of about six to eight months. In May 1960, Taylor wrote to express thanks to MacMillan for sending him a copy of the Anglican Book of Common Prayer published in 1744. "We are slowly building up a collection of early editions of the Bible and prayer books to which this will be a useful addition," he wrote.[15] By December 1960, MacMillan had furnished Taylor with a list of books on church history from The Publisher's List circulated from 83 Marylebone High Street in London. He asked Taylor to check the books he would like to have for the college library. Taylor gave instructions to his professor of church history:

> To mark the books: 1) that give particular attention to books published some fifteen years or more previously and which would not by then be available except through special dealers, and 2) that he should limit himself to some twelve or fifteen books.[16]

Taylor, careful not to over estimate MacMillan's generosity, wrote to him that he was uncertain as to how many books to check on his list. The total cost was £28.15. MacMillan wrote back and asked, "What others *do*

The Reverend Dr. William Stephens Taylor was principal of Union College at UBC, 1948–71.

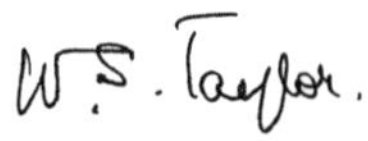

they want?" Clearly, this benefactor felt his commitment to the college had not been fully understood. He was inviting further opportunities to support its work through the donation of books.

MacMillan's generosity seemed to have no limits. Not long before Christmas 1961, he wrote an uncommon one-page letter to Taylor to accompany a rare and precious book, David Willson's *Impressions.* The book, so his letter claimed, was the first ever printed in Ontario. MacMillan understood that this book of reflections contained sermons preached by one of his ancestors, a leader in the Society of Friends, between 1790-1810. In his letter, MacMillan describes the rift over music in the Quaker Meeting House that led to a long-ago schism.[17] MacMillan, of course, was referring to David Willson, the founder of the Children of Peace mentioned earlier.

With David Willson's book came others: volumes of histories dealing with the Reformation, 19th century America, and the Crusades. A month earlier an even larger shipment of books arrived from MacMillan dealing with the histories of Ireland and England, social movements and social justice issues, and political associations. These subjects give us a glimpse into MacMillan's personal reading list and the books he chose for his own library.[18]

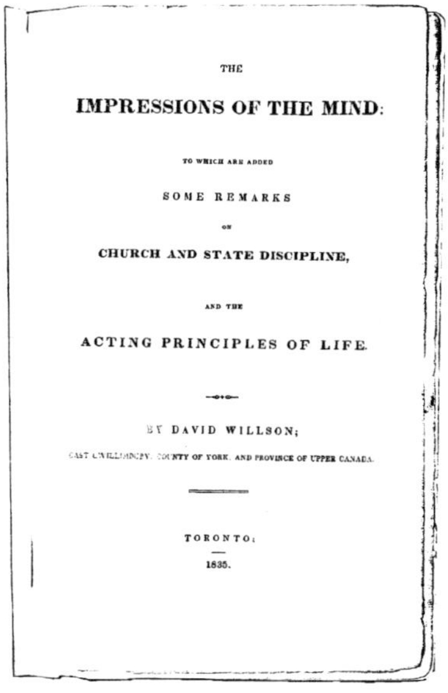

THE

IMPRESSIONS OF THE MIND:

TO WHICH ARE ADDED

SOME REMARKS

ON

CHURCH AND STATE DISCIPLINE,

AND THE

ACTING PRINCIPLES OF LIFE.

BY DAVID WILLSON;

COUNTY OF YORK, AND PROVINCE OF UPPER CANADA.

TORONTO:

1835.

Title page from David Willson's *Impressions of the Mind*

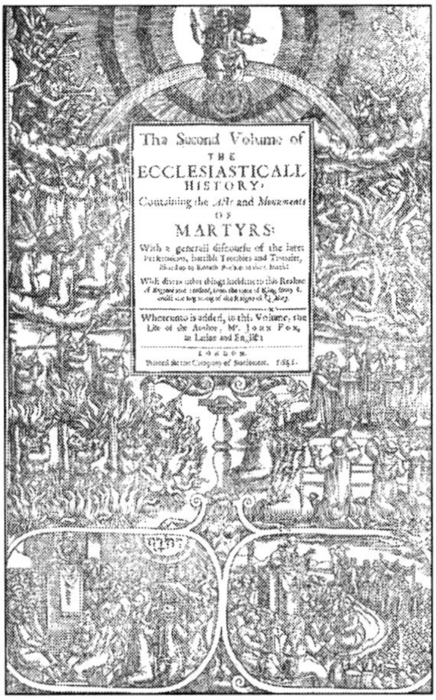

The Second Volume of

THE

ECCLESIASTICALL

HISTORY:

Containing the Acts and Monuments

OF

MARTYRS:

Whereunto is added, to this Volume, the Life of the Author, Mr. JOHN FOX, in Latine and English

LONDON.

Title Page; John Fox's *The Ecclesiasticall History of Martyrs*

Book of the month

MacMillan seemed to enjoy the opportunity to share some of his library with Union College. For a while, he sent at least a book each month. In December 1961, Taylor wrote to acknowledge some forty volumes he had received from MacMillan and he reassured MacMillan they would be well cared for and stored securely. Taylor drew attention to the particular need to care for the following books: Clarkson's three-volume *Portraiture of Quakerism*, Baler's *True Stories of New England Captives*, and the Reverend William Goode's book on the *Gifts of the Spirit.*[19] In late December 1962, MacMillan sent several copies of *History Today* for the reading room, and in early January 1963, he sent S. Gregorio Magno's *Imorali Del Pontefice S. Gregorio Magno*, and *Quakers in Science and Industry* by Arthur Raistrick.

Pastoral sensitivity

About the middle of February 1963, another carton of books arrived. Taylor's response was grateful, yet he also carefully pointed out that one of the books, Leibmann's *Peace of Mind*, was a duplicate of one already in the library. He asked permission to make the book available directly to the faculty and expressed his thanks for copies of the *Manchester Guardian Weekly* and *History Today*.

MacMillan drew Taylor's attention to some articles in the *Guardian* by S.G.F. Brandon on various aspects of the early church, particularly one on the Apostle Paul. He mentioned to Taylor his interest in one of these articles concerning the Jewish-gentile debates of the first-century church. Taylor responded that Dr. Brandon's distinctions "between St. Paul's beliefs and those of the Jewish Christian group were now pretty generally recognized by theological students nowadays."[20] However, he felt that Dr. Brandon might have overdrawn the distinction. He went on to describe his reasons for thinking so, and then concludes his letter:

> St. Paul has always interested me because of the great freedom with which he borrows ideas from any culture in which he is placed and uses them to communicate his Christian faith in ways that have meaning for his hearers. In trying to convey the meaning of the Cross, he borrows ideas from the Roman law courts, from Greek relations of friendship, from the relation of master and slave, from the practices of adopting children, and from the temple sacrifices. There is much we could learn from him in

> these days when there seems to be so much misunderstanding between the pulpit and the Press. Brandon seems to me to bring out this versatility of St. Paul's rather well. Thank you again for your continued interest in our library.[21]

Controversies plagued the first-century church as they came to grips with the implications of Jesus' teaching, and those of Paul and Peter. One controversy, recorded in *The Acts of the Apostles,*[22] focused on initiation into the church for Jews and/or gentiles. At one point, Peter, in a sermon, declared, "Can any one forbid water for baptizing these people?"

MacMillan and Taylor exchanged correspondence at a time when MacMillan was moving toward a deep reflection about his own life and death which may have coincided with the baptismal course underway with Fr. Damasus Payne in January 1962 and with the death of MacMillan's wife. It was a dark and sobering time for an aging gentleman. God moves in mysterious ways: what began as a casual comment from MacMillan led, over the years, to an exchange with Taylor concerning biblical issues, theology and modern scholarship, and now, the meaning of baptism.

Growing trust

Taylor's responses to MacMillan were immediate and sensitive. Through their correspondence, each man grew more open toward the other, and they began to find this exchange spiritually fulfilling. The relationship deepened to one of appreciation and trust mixed with mutual respect and admiration. In retrospect, each contact between the men seemed to be laying the groundwork for the major commitment MacMillan would one day make. Beginning as a donor of select rare books, MacMillan moved to become a principal benefactor giving one million dollars in 1965 to each of the two theological colleges on the UBC campus.

MacMillan's interest in the college was deeply rooted. By July 1962, he decided to give $5,000 annually for three years to assist Union College in appointing a director of development and recruitment, which was a first for any theological college in Canada. The funds for this new initiative came from the H.R. MacMillan Family Fund. His usual crisp and clear instructions to W.E. Baxter, Secretary of the MacMillan Family Fund Committee, were penciled on his letter, "Baxter, please put amongst my obligations."[23] Later that same year, MacMillan donated more rare books on the early work of the church in British Columbia to Union College. He included with them the three-volume *Book of Martyrs* (also known as *The Ecclesiasticall History of Martyrs*), written by John Fox in 1641 and originally published in 1554.

Practical projects

MacMillan rarely visited the two theological colleges at UBC. In January 1963, he accompanied VanDusen on a visit to Union College where he told Taylor that he would like to see a study conducted of a community such as Port Alberni, a Vancouver Island town where MacMillan had major investments. He wanted to research the ways in which people spent their leisure and holiday time and the relation of the churches to the problem of excessive drinking in the community.[24] Taylor did not forget MacMillan's comment about such a study and for eighteen months he worked on the concept of a project before writing to MacMillan to suggest a possible plan. Taylor did not miss a beat as he concluded his letter of application and request for financial help:

> I would be glad of your advice about these things, particularly as to whether you are interested in having us go ahead with the project, and whether it is feasible to have the students employed, for awhile, in the Mill at Port Alberni.[25]

Taylor kept his concern for his Union College students close to his heart – especially when the end of term was imminent. To pay next term's fee, Taylor's students needed summer employment and MacMillan appreciated his commitment to his aspiring student ministers. MacMillan agreed to fund Taylor's project and by April 27, 1964,, he instructed Peter Kaye, the Executive Director of the Vancouver Foundation, to: "pls arrange it (*the Alberni project*) – what will it cost? I will pay one way or another. *HRM.*"[26]

For many years, MacMillan sent turkeys to friends, business associates, and to those in need at Christmas. Taylor and his wife were among those who received a handsome bird, which they shared with the college faculty of the at a festive meal. Taylor took care to send his thanks "for the very fine turkey" and to report on the party he and his wife had hosted. He shared with MacMillan some details of his own personal and family life which added some human interest to his usual news concern-

DESIDERII ERASMI
ROTERODAMI
OPERA OMNIA
EMENDATIORA ET AVCTIORA,
AD OPTIMAS EDITIONES PRAECIPVE QVAS IPSE ERASMVS
POSTREMO CVRAVIT SVMMA FIDE EXACTA,
DOCTORVMQVE VIRORVM NOTIS ILLVSTRATA.
TOMVS PRIMVS,
QVI CONTINET
QVAE AD INSTITVTIONEM LITERARVM SPECTANT,
VT VERSA PAGINA DOCEBIT.
LVGDVNI BATAVORVM,
Curâ & impensis PETRI VANDER AA, M DCC III.
Cum speciali Privilegio Illustr. ac Praepotent. Ordd. Holl. & West-Frisiae.

Erasmus Title Page *Opera Omnia.*

ing the church.[27] He never failed to express his appreciation, and that of his staff, for the annual gift.

In October 1964, Taylor wrote to MacMillan to request help acquiring research materials for the College.[28] Nothing was closer to MacMillan's heart than research and, to that end, he had already donated the ten-volume *Opera Omnia* by Erasmus. Next, Taylor sought financial help to build a significant resource on the subject of the Protestant Reformation. He asked MacMillan to purchase a forthcoming reprint of the complete works of John Calvin (fifty-nine volumes) – a 1964 reprint of the 1863-1900 edition and released as a complete set for a cost of $1,225.00. Taylor said he was hesitant in making the request because of MacMillan's "recent very generous contribution to the Library. On the other hand, I realize that you might be disappointed if you were not informed of the fact that the reprint of Calvin's complete works is about to appear."[29] MacMillan simply instructed his secretary to "OK" the request and by December 31, 1964 Taylor had signed the receipt and acknowledged the gift.

Closer relationships

The correspondence between benefactor and principal reveals qualities in Taylor's character and personality that impressed MacMillan. Taylor observed every administrative step – he was a man with an eye for detail who could have been an effective manager or executive officer in MacMillan's vast corporate empire. Taylor was consistently grateful, courteous, thorough, diligent, and efficient. In all of his correspondence, he chose his words carefully, measuring each idea and phrase – MacMillan noticed and appreciated these qualities. Taylor was not only a highly regarded academic; he was also a humble and compelling servant of God. His long tenure as principal of Union College (1949-71) gave him the opportunity to develop his ecumenical goals for theological education despite the denominational sensitivities and inordinate time and patience required to make progress in inter-church cooperation. It was Taylor's vision that Union College of British Columbia and The Anglican Theological College be united in the teaching of theology and in the preparation for ordained ministry. This dream attracted Taylor's chief benefactor and would be realized in a surprising way.

MacMillan's advisers were his long-time colleagues and trusted friends. Several of his closest associates were the chief executive officers of institutions that he trusted and in which he was personally invested. MacMillan, VanDusen (vice-president of MacMillan Bloedel and the chair of the Board of Governors of Union College), Norman MacKenzie (president, University of British Columbia), Peter Kaye (executive director, Vancouver Foundation and a member of the Board of Governors of the Anglican Theological College), and Godfrey Gower (the Anglican Archbishop) often lunched together at the Vancouver Club. Their meetings provided a close-knit network in which MacMillan was encouraged to support theological education. He was now ready to endorse a vision of theological education, research and training in British Columbia and to support it with a substantial financial endowment. The second phase of MacMillan's faithful philanthropy was moving to a new and unprecedented level.

Chapter Eight

A Time to Endow

Anyone in their 80th year of life and having retired from an active career is bound to reflect on what she or he has achieved. MacMillan knew he had changed the BC forest industry forever. He realized he was old and that he no longer had the energy required for the close management of his business empire. Nonetheless, he remained bright and alert, and was deeply interested in his family, investments, and the everyday dynamics of community life.

At the pinnacle of his business career, H.R. MacMillan owned an industrial complex that included a shipping line that carried MacMillan Bloedel lumber to world markets.

For MacMillan, 1965-75 was a time for things of the spirit, always present within him and now brought to the fore. His sense of responsibility for, and gratitude to, his community was increasingly expressed through a wide range of small- and medium-sized financial gifts. He was in command of considerable wealth and felt he must support aspects of the community that reflected his lifelong values.

Principal W.S. Taylor, and others, had gently nurtured MacMillan's renewed interest in the support of Christian leadership and the training of personnel for ordained ministry. MacMillan's approach to a new phase in his philanthropic career was similar to his long-established pattern of decision-making in the business world: he assessed needs; he saw opportunities; and took the necessary action. He decided to finance the two theological colleges at the University of British Columbia in a special way and gave instructions to his lawyer, Charles William Brazier, to prepare the papers.

A gift revealed

When the Executive Committee of the Board of Governors of the Anglican Theological College of British Columbia sat down to a meeting on the February 3, 1965, no one had anticipated anything special to occur from previous meetings. There was no printed agenda, and most governors thought they had come to nothing more than a regularly scheduled meeting. The one member chairing the meeting, however, had a significant item on the agenda. As the meeting unfolded, there was not the slightest inkling that a forthcoming announcement would change the college in dramatic ways.

The President of the Board of Governors, Bishop Godfrey Gower, led the executive through its routine business. The minutes taken contain the items on the agenda: the executive took attendance, noted regrets, and considered the previous meeting's minutes; they heard a resignation from the Board of Governors; listened to a report of a committee recommending honourary degrees; acted on a report from the Constitution and By-Laws Committee; and received two financial reports. The president had kept the news of the pending gift to himself, until he finally revealed that he had not been at liberty until that day to disclose to the governors that, "Mr. H.R. MacMillan wished to make a substantial gift to the college.[1] He had requested his solicitor, Mr. C.W. Brazier, Q.C., to enquire as to the most important need of the College at this time." The minutes contain the announcement:

Anglican Theological College, Chancellor Boulevard at UBC was built in 1927.

> Then read a memorandum[2] which he had prepared for submission to Mr. MacMillan, the purport of which emphasized the great need for a fund to provide for post graduate and advanced studies for the teaching staff; for fellowships to provide for the services of visiting lecturers and for short term refresher courses for clergy throughout the Province.
>
> Mr. MacMillan had instructed his Solicitor to prepare a Deed of Settlement conveying a gift of 30,000 ordinary shares of the capital stock of MacMillan, Bloedel and Powell River Limited (known as "MB & PR) establishing an endowment fund of approximately One Million Dollars.
>
> The President stated that no publicity was to be given to this gift, pending a press release which, if made, would be by the donor.[3]

Bishop Gower then explained that he had hoped to call the executive earlier, but that action had to be taken almost immediately because MacMillan wanted the papers to be completed before he left the city. Five members of the executive committee met to examine the proposed deed of settlement and "see it signed and completed." Principal John Blewett of the Anglican Theological College immediately expressed his great satisfaction at this most timely gift and much favourable comment followed. The committee ratified the action of the president and resolved that an expression of gratitude be sent to the donor.[4]

The Rt. Rev. Godfrey Gower, Bishop of New Westminster and president of the Board of Governors of the Anglican Theological College at UBC frequently had lunch with H.R. MacMillan at the Vancouver Club.

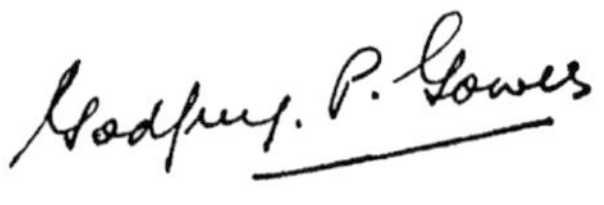

Union College responds

When the Union College of British Columbia was notified of the pending endowment, it responded slightly differently from the Anglican College. A special meeting of the Executive and Finance Committee of the Board of Governors was called for Saturday January 30, 1965, at which ten members were present. Mr. Guy Flavelle chaired the meeting as Mr. W.J. VanDusen's term as President of the Board of Governors had concluded in 1964. The endowment was the meeting's only business, and the letter to Dr. Taylor was read. The minutes record the event:

Deed of Gift from H.R. MacMillan for Endowment.

> Mr. Walden read a letter (from C.W. Brazier) saying that Dr. H.R. MacMillan had executed a Deed of Settlement transferring 30,000 shares of capital stock in MacMillan, Bloedel and Powell River Ltd., to be held in trust as an endowment fund for Union College of British Columbia, the approximate value of the shares being one million dollars.
>
> It was noted that the objectives outlined in the Deed of Settlement corresponded very closely with the objectives which the Board and Senate had themselves laid down for the future development of the college.
>
> It was carried that the College accept the gift with gratitude; and that a letter be sent to Dr. H.R. MacMillan expressing the thanks of the Governors for the gift, and their appreciation of the terms in which it had been made.
>
> It was noted that there must be complete silence about this gift until a formal press release was issued.[5]

Union College, the seminary of the United Church of Canada located on Iona Drive at UBC was built in 1927. In 1960, H.R. MacMillan gave the stone facing for the central tower. In 2004, the building was totally gutted and its rebuilding was completed in 2006.

At the conclusion of the meeting, Principal Taylor could not wait to write to MacMillan. With characteristic efficiency, it took him less than twenty-four hours to acknowledge the gift and to assure MacMillan that strict confidentiality would be kept in regards to the endowment. He included expressions of gratitude from the Executive and Finance Committee for the gift, as well as its appreciation for the thoughtful way in which the terms of the gift were stated. In his letter, Taylor noted that the objectives outlined in the Deed of Settlement* corresponded with the hopes and plans that the governors held for the future development of the college.

Silence kept

No announcement was made in the press regarding the endowments to either college. Further, no statement was sent to the Anglican or United Churches from the two colleges that suggested in any way that the gift was forthcoming. MacMillan himself broke the news to the press and was out of the city when it appeared in Vancouver newspapers. On February 10, 1965, *The Vancouver Sun* described the two gifts to the theological colleges, along with $3.3 million MacMillan donated to provide

* See Appendix II

The University of British Columbia was the recipient of the greatest portion of the one hundred million dollars H.R. MacMillan directed to charitable and educational organizations. He gave mulitple numbers of gifts and scholarships to UBC each year over a period of forty-five years.

graduate scholarships at UBC and another $3 million for the UBC library, for a total of $8.3 million. On the same day, *The Province* quoted UBC President, John B. Macdonald as saying, "These gifts are the most generous ever made for graduate education in Canada."[6]

The endowments were so shrouded in secrecy that even MacMillan's family was unaware the gifts had been made. The awards to the two theological colleges came out of MacMillan's personal resources and the postgraduate fellowships at UBC were donated by the MacMillan Family Fund, a private foundation. Following the public announcement by MacMillan, the two colleges announced the details surrounding the gift to their respective constituencies. In his letter to the bishops and clergy dated February 25, 1965, Bishop Gower said he was anxious that the clergy be "properly informed as to the nature and purpose of the gift to the College made by Mr. H.R. MacMillan." Bishop Gower expressed gratitude for the gift and wrote, "The benefit to the clergy throughout the province will be incalculable and the whole life of the Church stimulated and strengthened."[7]

One of the first people to commend MacMillan for his gifts was Dr. Norman MacKenzie, recently retired president of UBC and a close friend of MacMillan:

> I am particularly pleased that two million dollars goes to the Union College and Anglican Theological College. While the churches are having a difficult time, I would hate to think what our society would be like without them and the influence they have had, and still have, in the area of morals, standards and responsibility. I know most of the staff in the two colleges and they seem to be first rate and are doing very good work.[8]

Norman A.M. MacKenzie

Dr. Norman "Larry" MacKenzie (left) with H.R. MacMillan at a formal dinner party at UBC Faculty Club on September 29, 1961, given by MacMillan's appreciative guests on his yacht *Marijean*.

The former president was responding not only to the news release concerning the endowments but also to the much larger gift to UBC ($8.3 million) that MacMillan announced at the same time. MacKenzie also believed that these gifts were the largest ever made to graduate education in Canada – an endowment of $3.3 million provided for post-graduate fellowships for some sixty-three Ph.D. candidates and a gift of $3 million enabled the library to increase the numbers of books from 700,000 to two million. UBC now had one of the most heavily endowed university libraries in North America.[9] Within the year, MacMillan donated a further $750,000 for graduate studies in the Institute of Fisheries and a further $670,000 for a ten-year book fund for the library.

The terms of the endowment

It appears that VanDusen, Taylor, and Gower each contributed to the terms of the settlement laid down by MacMillan for the theological colleges. Although he did not wish to be identified, it was VanDusen who likely encouraged MacMillan in his endowment to both Colleges. VanDusen completed his term as chair of the Board of Governors on

November 5, 1964, and although he was highly influential in encouraging MacMillan to present the endowments, he had detached himself from the gifts and the public announcement.

Principal Taylor denied he knew the endowment was forthcoming. Furthermore, he said he did not know the Anglican College was to receive an additional gift. "It came like a bolt out of the blue," he said.[10] His comment is quite consistent with his character – he had focused on MacMillan's inner journey by becoming his spiritual mentor but was unconcerned with MacMillan's philanthropic aspirations. Taylor modestly claimed he had played but a small part in the process that led to the endowment.

Many years later Taylor reflected on MacMillan's growing commitment to Union College, saying that he came to understand the significance of MacMillan's annual Christmas gift of a turkey. Taylor felt that the gift represented MacMillan's deepening commitment to the college and theological education. "Sometimes a little thing like that," said Taylor, "shows a real interest, and can be as significant as a large gift."[11]

Precisely who or what process had shaped the terms of the endowment remain unknown. Bishop Gower has claimed he prepared a memo for MacMillan to use as a guide, but this document cannot be found. Taylor played only a small part. It is probable that all three close confidants of MacMillan (Gower, Taylor, and VanDusen) had input into the terms of the endowment. What is clear is that its explicit terms set forth by MacMillan closely reflect his understanding of what is required to prepare first-class leaders in all fields. Perhaps he alone had shaped the terms, recalling how he had been the recipient of financial aid as a student. In MacMillan's view, leaders in all professions can never be too highly trained and for him, this belief was especially relevant for those entrusted with the care and nurture of others in matters of the spirit. As an undergraduate student, MacMillan had been inspired by visiting lecturers, and he had gone on to graduate school. He had the opportunity to pursue his professional interests in many parts of the world, and he endorsed the idea that bold and creative initiatives often benefited the wider community also. His terms were clear:

TO DEVELOP A PROGRAMME OF RESEARCH, SCHOLARSHIP, AND TRAINING, IN PARTICULAR,

1. to afford members of the teaching staff of the beneficiary opportunity for advanced study in larger centres;

2. to establish fellowships in order to attract the services of visiting lecturers expert in their field;
3. to offer post-graduate courses;
4. to afford opportunity for studies abroad for those showing competence in special fields;
5. to offer short term refresher courses for the clergy throughout the province of British Columbia, and
6. to undertake such other projects of a similar nature as the governing body of the beneficiary may decide upon from time to time.

It was MacMillan's specific wish that the advanced studies and training undertaken in the theological colleges would include the greatest possible degree of cooperation with other denominations, including the exchange of personnel and the results of research and study.

The two colleges immediately named their respective committees to bring recommendations to their Executive Committees and Boards of Governors concerning the administration and disbursement of the funds. Union College appointed three members with the power to add more if necessary, and the Anglican College appointed "not more than seven members" to deal with all matters covering the MacMillan gift.

Two colleges unite

The timing of the endowments to the two theological colleges must also be understood in the context of a larger ecumenical conversation then underway between the Anglican Church and the United Church of Canada. In 1965, the union of the two churches was under serious consideration at the national level, and a plan was underway for the two denominations to meet that year. An air of optimism existed that their respective national bodies, the General Synod and the General Council, would approve the union. Furthermore, there was already a degree of cooperation and joint efforts between the two UBC colleges.

When the colleges learned of the identical gifts they were to receive, they agreed to cooperate wherever possible. By June, a joint meeting of the respective MacMillan Fund Committees met to discuss proposals for cooperation and, by December, the colleges agreed to the appointment of a joint librarian, The Reverend R.M. Clark. The colleges joined together six years later to form the Vancouver School of Theology (VST) and the two endowments merged into one.[12]

The hundreds of awards made and distributed throughout British Columbia from the H.R. MacMillan Fund for Theological Education reflect the VST's original terms of reference and its subsequent priorities.

Six categories exist to guide the committee in its disbursements of the MacMillan Fund Committee: Non-Faculty Clergy, Faculty Leave/Conferences, Visiting Professorships/Sessional Lecturers, College Library, Pastoral Education, and Continuing Education. Between 1965 and 2005, the MacMillan Fund Committee distributed more than six million dollars. Hundreds of clergy have benefited from short and long-term courses and pursuing higher degrees. Interdisciplinary pursuits have also been encouraged as extensions of the Christian ministry and the library has benefited enormously. The MacMillan Fund Committee has received hundreds of letters from grant recipients expressing appreciation for the financial support they received.

VST

Logo of Vancouver School Of Theology

A range of benefits

In addition, the endowment fund has supported other aspects of the Vancouver School of Theology, which have included support for visiting professors and lecturers, the Native Ministries Consortium, the annual summer school programs, and a variety of ecumenical initiatives in the church and ministry programs. These latter programs are known as the Chalmers Institute for Continuing Education and Congregational Development, so named after the generous VST's endowment from Chalmers United Church where the MacMillan family was a member for many years.

MacMillan's commitment to funding theological education has enormously strengthened the VST's work serving parishes and communities far beyond its own students and alumni. In keeping with MacMillan's vision of greater denominational cooperation, the college's library and faculty work increasingly closer with new church-affiliated colleges on the UBC campus such as Regent College, Carey Hall, Saint Andrew's

Hall, and St. Mark's Hall. The VST library, now named The H.R. MacMillan Theological Library is a part of the larger library system of the University of British Columbia.

The H.R. MacMillan Theological Library has expanded its services by supporting distance education students from St. Stephen's Theological College in Edmonton. The Chalmers Institute for Continuing Education and Congregational Development Studies directly benefits from MacMillan's farsighted gift in the form of a yearly allocation of funds to hire staff for related programs. Furthermore, faculty members belonging to this VST extension assist with programs for clergy and laity from other parts of Canada and beyond on sabbaticals at the School. VST, known among theological colleges and seminaries for its self-directed studies program, enables clergy to pursue specific areas of interest for long or short periods of study.

Furthermore, the Chalmers Institute, in addition to developing a program in Christian spirituality, has developed the Canadian Studies Certificate Program for Japanese students. The Disciple Bible Study Program and customized computer workshops are other, newer aspects of the VST's expanding list of learning resources for the church in British Columbia.

Since the endowment was established in 1965, MacMillan's generous gift and his vision for theological education have been an inspiration to hundreds of members of the clergy. Although the needs of theological education far exceed the VST's ability to respond within the limits of its the present resources, it is nonetheless profoundly grateful to MacMillan for his vision and commitment, which are so generously expressed in and through his splendid endowment. In retrospect, what makes this gift of immense significance is its ability to enable the churches to move confidently and creatively through four decades of great change and turmoil.

Chapter Nine

A Time to Dream

H.R. MacMillan did not believe that business leaders should hide in secrecy behind a meaningless company name. For this reason, he was clearly identified in the name of his first business, The **H.R. MacMillan** Export Company Limited and that of his last enterprise, **MacMillan,** Bloedel Company Limited. "I don't believe in people in business hiding in the anonymity of a company name that doesn't say who is there,"[1] he said. The up-front exposure of MacMillan's name with his company was in keeping with the British business tradition of clearly identifying a personal commitment and accountability to the public. He held the motto of the British Merchant Traders Association in high esteem, "Make your name and make it a good name."[2]

He believed in transparency, which he proved by branding his personal initials, **HRM**, on every single piece of lumber he produced! At the time of MacMillan's death, Cyril A. Woods of Oregon wrote a letter of condolence to MacMillan's daughters:

> Some day fifty or seventy-five years from now, some old building will be torn down in Shanghai, Johannesburg, Calcutta, London or Buenos Aires, and timbers will be exposed bearing the letters "HRM" boldly stenciled. This shows how far and how long the influence of his career will be evident. (I'm sure this will happen because I have shipped millions of feet of lumber from the Portland office alone with these letters on all of it).[3]

In this use of his initials in such a way, there is in MacMillan a touch of the student of the classics. Given his nature and his interest in the arts and literature, it was perhaps not surprising – his name was important to him. His companies bore his surname, and he christened his ships with his family's name – *Harmac* (**Har**vey **Mac**Millan) and *Marijean* (**Mar**ion and **Jean**) his daughters. Likewise, he named the company newsletter the *Harmac News*. "It is a good principle," he said, "to make your name good and then make it known."[4]

MacMillan remembered

Thirty years after his death, MacMillan is still a household name remembered by British Columbians. For some he was simply a person of great wealth, a rich lumber baron; others remember him with appreciation for establishing British Columbia's early forest management and harvesting policies. He is also remembered as the first person to develop and sustain the industry based on a significant export product from BC. *The Times* of London echoes these words: "More than any other man he was credited with establishing British Columbia as a major supplier of forest products to the world."[5] *The Canadian Encyclopedia* describes him as an entrepreneur, an aggressive businessman.[6] For others, he was a scholar, a conservationist, and an environmentalist ahead of his time. Some remember him for the titles and honours various societies and universities bestowed upon him: C.B.E. (Commander of the British Empire); B.S.A. (Bachelor of Science in Agriculture, Toronto); M.F. (Master in Forestry, Yale); C.C. (Companion of the Order of Canada); L.L.D. (Doctor of Laws, Toronto); D.Sc. (Doctor of Sciences, UBC); and F.R.C.A. (The Royal Society for the Encouragement of the Arts, Manufacturers and Commerce).

Support of the arts

The previous chapters explored some of MacMillan's many gifts to educational, social, and religious institutions that were on his list of causes worthy of his financial support. Furthermore, he felt that the arts play an important role in the life of the community. MacMillan collected works of art, particularly those of the Canadian Group of Seven, and his home was the gracious setting for their artistic renderings of Canada. Lawren Stewart Harris, one of the group and the son of a founder of Massey Harris, the farm equipment company, discussed with MacMillan the need to support an extension to the Vancouver Art Gallery that would house a collection of Emily Carr paintings. MacMillan made the first donation of $10,000 to the campaign for capital funds, and he asked his colleagues to match this gift. This initiative raised $50,000 for the Gallery,[7] which in 1947 represented a substantial amount, and again led by example.

MacMillan's friends included artist A.Y. Jackson, a member of the Group of Seven and they corresponded occasionally. When MacMillan was eighty-six years old, he wrote a note to Jackson saying how much he was enjoying the four Jackson paintings hanging in his living room. Although he was far too frail to visit Jackson, who lived at the McMichael Gallery near Kleinberg, Ontario, they corresponded and were able to remain close friends despite the great distance between them.

A year after MacMillan moved to Vancouver in 1918, he joined the Vancouver Club. It appears that the Vancouver Club served his interests well and, when the Club began to acquire works of art, MacMillan responded. Frank Auger, a member of the Picture Acceptance Committee, was invited to MacMillan's home for cocktails and to discuss a possible gift. MacMillan invited him to choose something from his collection and Auger chose a painting by Diego Rivera. Later, however, MacMillan revealed that some of his family was not pleased with his decision to donate this painting, so instead he provided the Vancouver Club with an Emily Carr painting, *The Crooked Staircase,* which is probably the most valuable one in the club's collection.[8]

Visions and dreams

MacMillan's philanthropy included more than simply disbursing his wealth to a variety of institutions. His priorities for support reflected his dreams and hopes for those institutions that fostered and enriched community life including libraries which received his generous support. MacMillan had a fascination with architecture and history – particularly, the history of England – and requests to support projects that celebrated history, both in written form and in art, fired his imagination repeatedly.

Following World War II, the oldest church in the City of London, All Hallows by the Tower, captured MacMillan's interest. The church's Saxon beginnings date back to 675 A.D., and it has a colourful history as it has been rebuilt and restored a number of times since then. For example, in 1650 it was blown apart by the explosion of twenty-seven barrels of gun power, and it was fully restored by 1658. When William the Conqueror ordered the construction of the Tower of London following his successful invasion in 1066, All Hallows, located near the new tower, was already 400 years old. Subsequent dire and gory goings-on at the Tower of London next door meant that many headless bodies of those who displeased the monarchy were sent in All Hallows churchyard for burial, including that the Archbishop of Canterbury, William Laud, who was beheaded at the tower and buried at All Hallows in 1645.

Miraculously, the Great Fire of London in 1666 failed to consume All Hallows. However, World War II did not leave the church unscathed: Nazi bombs destroyed the church during the 1940s blitz and the brick tower and walls were all that remained standing after the raid. The parish rector and wardens decided to rebuild the old church, and in 1948, the Queen oversaw the laying of a new cornerstone. In 1949, Sir Ion Hamilton Benn, a friend of H.R. MacMillan, asked him to provide long timbers from British Columbia to rebuild a spire on top of the still-standing tower. According to The Reverend P.B. Clayton, the rector, the timbers came "forthwith" and the spire was completed by 1958.

H.R. MacMillan supplied the interior beams for the spire of All Hallows by the Tower, Dockside, London. The rector called it Ion's Spire, in honour of the man who led the financial campaign and who appealed to MacMillan for the great timbers in 1950.

MacMillan's interest in this project was possibly stimulated by the historically significant Quaker link that All Hallows had with Admiral Sir William Penn. The Admiral's son, William, founded the Quaker colony in Pennsylvania. He was born in 1644 and baptized in All Hallows Church the same year. Some twenty years later, Penn saved All Hallows from the Great Fire of London by having the surrounding buildings demolished to create firebreaks. Was it this Quaker connection, more than any other, which prompted MacMillan's generous response to assist in the rebuilding of "Ion's Spire on the church tower of All Hallows by the Tower of London," as it was called by the rector in a brief note to MacMillan in 1960?[9]

The Wren connection

Another visionary project drew both MacMillan's attention and resources. This dream had its genesis over a period of years during which MacMillan was in frequent contact with Dr. William C. Gibson, professor of the history of medicine at UBC. They were close friends and had worked together on a variety of projects, including MacMillan's search for rare books and the procurement and placement of what Gibson described in an interview with the author as "The great Chinese tapestry illustrating the work of Dr. Norman Bethune." Not only was MacMillan interested in the unfolding drama of medical advances, he easily identified with the Bethune project for the simple reason that both he and

Dr. William C. Gibson, Faculty of Medicine at UBC and close friend of, and visionary with, H.R. MacMillan

The H.R. MacMillan home, surrounded by spacious lawns and gardens, was located in the Shaughnessy area of Vancouver. The house, now replaced by smaller units, is where MacMillan and Gibson met on Sunday afternoons to discuss their plan to acquire two pieces of Wren architecture for Vancouver.

Bethune had contracted and recovered from pulmonary tuberculosis early in their life. Furthermore, Bethune, as a surgeon, devoted much of his life treating tuberculosis victims in his home province of Quebec. That reason alone was enough for MacMillan to purchase the tapestry and present it to UBC. Another gift he made to the UBC medical school was a series of fifty display panels placed in the Woodward library to exhibit important medical discoveries and achievements.

In the latter years of MacMillan's life, he and Gibson visited each other every Sunday afternoon. They discussed books they had read, and they shared their thoughts and feelings through correspondence. In the 1960s and 1970s, their interests focused on subjects of a religious nature. MacMillan stimulated Gibson's interest in the Quakers –particularly their contributions to the advancement of medicine. Gibson engaged himself in a search for literature of the Society of Friends, and he used some of his findings in an address to the University College Hospital in London during his Osler Club lecture in November 1967.[10]

Gibson was anxious that MacMillan read *Jesus as Teacher,* a book published in 1935 and written by a University of Toronto graduate Henry Burton Sharman, a Quaker. Gibson himself read this book every Easter.

Like MacMillan, Sharman had links with the University of Toronto where he was a leader of young people's Bible study groups in the early 20th century. Sharman and Gibson knew each other personally as the latter was the resident doctor at Sharman's summer seminar at Lake Minnesing in Algonquin Park in 1934. Sharman was a biblical scholar trained at both the University Toronto and the University of Chicago and was particularly interested in the internal contradictions of the New Testament. According to Gibson, Sharman was under the impression that some "fast work" was done on the New Testament writings "with scissors and glue." Sharman was concerned today's version of the New Testament is not a true reflection of the original writings. MacMillan, interested in the point of view that Sharman presented, read the book and he wrote to Gibson saying that he was anxious to discuss it with him.

There is no record as to how the conversation unfolded between MacMillan and Gibson but they continued to share other interests. For example, both were fascinated by the achievements of Sir Christopher Wren, the famed 17th-century scientist and architect of St. Paul's Cathedral in London. Wren's biomedical discoveries were entirely new to MacMillan, and he asked Gibson searching questions about Wren's work on blood transfusion in dogs, which he undertook while in his mid-twenties. In 1656, Wren injected a warm solution of "opium in sack" into the bloodstream of a large dog.[11] "To MacMillan's widening eyes," Gibson described how Robert Boyle, the chemist, noticed that, "The dog began to nod with his head, and falter and reel in pace. After some time he began to come to himself again."[12]

Although MacMillan expressed concern for the animal used in the experiment, he was subsequently relieved to learn that, "The dog grew fat and was stolen from its owner," and a smile lit up his face.

The great church venture

In the early 1970s, Gibson and MacMillan explored in earnest a special project that focused on two of the more than fifty architectural gems in the City of London designed by Sir Christopher Wren following the Great Fire of 1666. Gibson called this project, "our great church venture," and he hoped it would be "worth all the effort."[13]

Their vision was to bring two fine examples of Wren's architectural masterpieces to Vancouver and place them in prominent locations for all to enjoy.[14] "The very idea brought tears to MacMillan's eyes," said Gibson.

The tower of St.Mary's Somerset near the docks of London was first proposed to be relocated to Christ Church Cathedral on Georgia Street at Burrard in Vancouver, BC.

The buildings in question were among those that had remained standing when the World War II bombing blitz of London came to an end. Some, who considered the ruins to stand in the way of London's post-war urban development, thought they should be disposed of. MacMillan and Gibson proposed that the eighty-four-foot tower of St. Mary's Somerset* on Upper Thames Street, near London's docks, be relocated beside Christ Church Cathedral at Burrard and Georgia Streets in Vancouver. Further, they suggested that the tower of Christ Church on Newgate Street, the Greyfriars Church, be erected at the north end of Main Mall at UBC to overlook Howe Sound.

Early in the project, Gibson and MacMillan realized that it might take years to achieve their dream, but they remained undaunted in their self-imposed assignment. Their 1971 proposal to the Vancouver Foundation seeking financial assistance, included a reference to a Wren-built

* Not only did this tower survive the blitz, it also remained standing after the nave was ordered demolished by Act of Parliament in 1872.

The first plan for the campus of UBC designed by architects Sharpe and Thompson. The latter is the person MacMillan contracted to build the library furnishings for Union College in 1927.

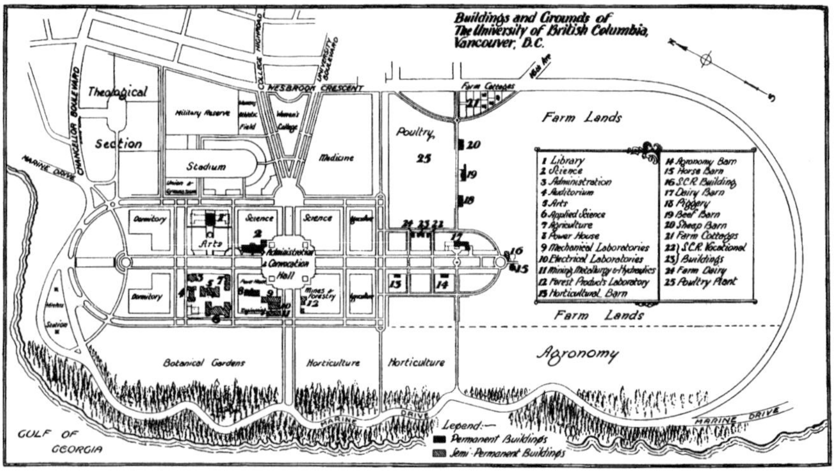

A bird's-eye view of the 1925 UBC campus plan shows the proposed location of buildings, including a large section set aside for theological studies. The north end of Main Mall is where MacMillan and Gibson proposed that Sir Christopher Wren's Greyfriars spire be located.

church that had been re-erected on the campus of Westminster College in Fulton, Missouri, in which Sir Winston Churchill had delivered his famous speech on freedom in 1946. The two visionaries argued that, "If a bombed-out Wren church could be relocated to [the] USA, was it wishful thinking, utter dreaming, that the same might be achieved for Vancouver and perhaps at UBC?"

The MacMillan-Gibson proposal had two objectives: to erect a Wren spire at UBC, and to place another in downtown Vancouver. The first church tower-spire selected was that of Christ Church, also known as Greyfriars Church on Newgate Street, near the east end of St. Paul's Cathedral in London. With its still-intact spire and three walls sheltering the remains of its burnt-out nave, it needed new walls and a roof. The restored church would serve as a chapel for a multi-denominational congregation of the university community. The proposal to the Vancouver Foundation stated it would be placed at the north end of Main Mall on the UBC campus. It would "front on the road leading from Chancellor Boulevard to the present Faculty Club with the body of the church running northward, its northerly wall to be largely glass so that the congregation could contemplate the view." In the immediate foreground would be the rose garden and, from the pews inside the chapel, the vista would include the waters of the Strait of Georgia, Howe Sound, and the snow-capped mountain of the north shore.

The estimated cost of the project was one million dollars plus transportation and the cost of reconstruction at UBC. The completed project would reflect the vision of the original design of the UBC campus as prepared by architects G.L Thornton Sharp and Charles J. Thompson in June 1914. (This was the same Charles Thompson whom MacMillan engaged to design and build the library furnishings of Union College in 1921.) The architects' plan featured a large group of theological colleges and seminaries to be located in the northeast corner of campus overlooking Howe Sound.

The second objective of the MacMillan-Gibson dream was the relocation of a Wren spire or tower to Christ Church Cathedral in downtown Vancouver. Gibson, in particular, immersed himself in the project, and he kept MacMillan informed of its developments in his correspondence. Both men were aware that the project might run into opposition because at the time another plan was unfolding for the use of the land occupied by Christ Church Cathedral. That vision included a high-rise business block on the corner of Burrard and Georgia Streets that would house the Cathedral congregation below ground level, and that

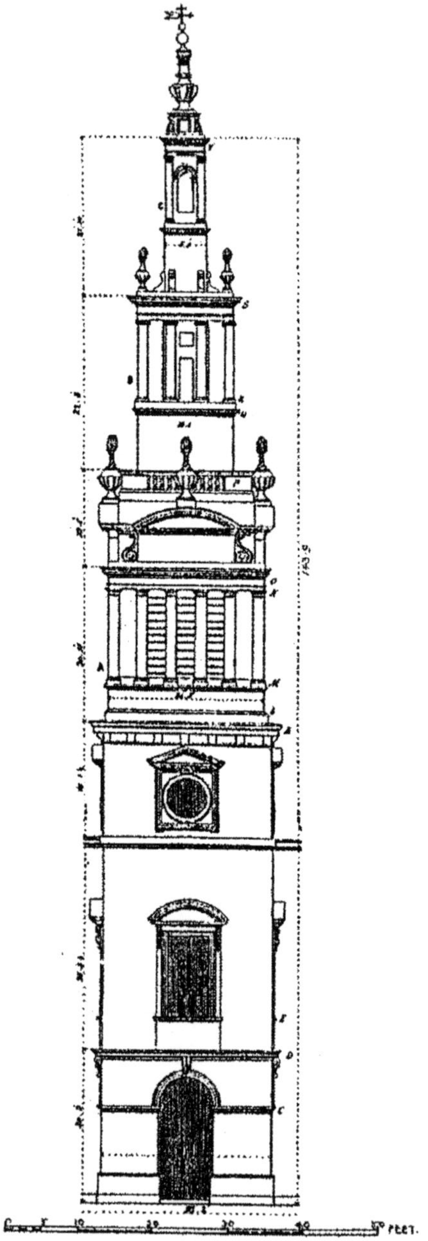

Wren's plan for the spire of Christ Church, Greyfriars, Newgate, London. It was proposed that the spire and nave be moved to UBC but, when that failed, the plan was to place the spire at Christ Church Cathedral at the corner of Burrard and Georgia Streets in Vancouver.

involved the demolition of the existing building (built 1894-95). Although the Cathedral congregation, after several well-attended vestry meetings, (including the whole congregation), approved this plan, Gibson and MacMillan did not support its vision. They wanted to see the Cathedral preserved and to have a Wren tower or spire placed at its southwest corner to serve as the entrance to the existing building. Their plan was consistent with the vision of C.O. Wickenden, the architect who drew the plans for Christ Church Cathedral in 1888 that included provision for a spire.

MacMillan began to finance the necessary research, and he committed to exploring arrangements for the shipping of a spire or tower from London to Vancouver. Gibson went to London to begin the negotiations with Church authorities, and on May 7, 1972, he sent a post card to MacMillan:

> *Dear HR,*
> *The destruction of Christ Church Cathedral in Vancouver angers me and it will Mountbatten when I tell him. Keep watch over it until I return.*
>
> *Bill*

While in London, Gibson dedicated himself to studying Wren's work, and he prided himself on learning to "recognize Wren's handwriting from hours spent pouring over his documents in the British Museum and the British Institute of Architects at Portland Hall."[15] As far as Gibson was concerned, "Wren's architecture is the glue that holds London together as a community," and he wanted an example of it in Vancouver.

As the years unfolded during this project, MacMillan's energies began to flag. Now in his late eighties, he began to express some hesitancy about it – for him, the dream seemed to be fading. UBC did not approve the relocation of the Greyfriars church and spire to Main Mall at the university. Furthermore, the City of Vancouver ruled in favour of the preservation of the existing cathedral building, and the city council was considering the construction of a major business development directly adjacent to it. In due course, the cathedral officers and the developer signed an agreement respecting the air rights above the cathedral, which included a provision that a spire could be added at any time in the future provided such a possibility did not exceed the footprint and vision of the original architect's blueprints. These blueprints allowed for a spire or tower, and Gibson began to work on this possibility immediately and he made plans to meet with British investors.

With the air-rights agreement in place, a renewed energy emerged regarding the possibility of relocating a British architectural jewel to the Cathedral site in Vancouver. Gibson was determined to achieve the dream. He and MacMillan both preferred the more graceful and elegant Greyfriars spire (at one-hundred-and-forty feet in height plus an eighteen-foot finial) to that of St. Mary's tower, and they suggested that the Greyfriars spire be located at Christ Church Cathedral. The Greyfriars spire would reach the height of the Panorama Room located in the Hotel Vancouver across the street from the cathedral. Failing this, another site might be found in Vancouver's Civic Square, as Gibson called the area known as Victory Square. Their single-minded goal was to have an example of Wren's art placed in Vancouver.

Gibson went in London to meet with decision makers. While at the Royal College of Physicians he sent a postcard to MacMillan on January 13, 1975:

> "Saw Greyfriars Tower – still intact. Hope we can take it home.
> Bill Gibson"[16]

The above corner of the Christ Church Cathedral facing Georgia Street and the Vancouver Hotel was the proposed location for Christopher Wren's Greyfriars spire.
Credit: A 1909 postcard from the Archives of Christ Church Cathedral, Vancouver

Advisors and supporters

Gibson enlisted the help of Sir Michael Perrin, an eminent scientist in England and a fellow Oxonian friend. Born in Victoria, BC, he was the son of a former bishop of the Anglican Diocese of British Columbia, and was married to the daughter of a bishop. He was well connected to the English scientific and ecclesiastical establishment. Gibson thought that Perrin's contact with the Bishop of London would prove useful in gaining permission to remove the 164-foot spire from London.

Now eighty-nine years old, MacMillan directed his limited energies to his personal needs, "my job," as he called it. Only a few friends visited him from time to time in his main floor suite of his home, a place he rarely left. Gibson knew that time was not on MacMillan's side in having their dream fulfilled.

By July 1975, the possibility of ever realizing their dream was diminished as the London decision makers did not approve of it. Gibson reported to MacMillan that Sir Michael Perrin had advised him that the required unanimity had not been reached, and that the Greyfriars spire, or any other, was considered too important to leave England.

Although MacMillan and Gibson discussed importing other Wren achievements, MacMillan's energy and personal commitment to the project were now expended. Their schemes, which were typical of MacMillan's imagination, vision and responsibility back to the community, were now set aside.

Set an example

MacMillan avoided personal publicity as much as possible. "Greatness," he said, "is measured by what you do for society. I haven't done much except to the degree where the creation of employment is considered a contribution. If I hadn't done it, somebody else would have. All I did was make some money."[17] In this respect, he could be heard expounding on his personal political philosophy as he addresses the 1950 spring convocation of the University of British Columbia, at which he received a Doctor of Science degree, *Honoris Causa*:

> My message to you in the graduating class, and I urge its importance, is that, in leaving University, you resolve to give to British Columbia as well as receive from it. That you study her problems and work for their solution, and that you regard the public business as your business. That you

examine carefully and critically those proposed solutions to public problems that emphasize the duty of the state to the individual rather than –what is much more important – the individual's responsibility to society and to himself. Possibly you should not be satisfied that there are only two UBC graduates in the British Columbia legislature.

The challenge is yours – to build upon the heritage of the past with courage, realism and unselfishness.[18]

Chapter Ten

A Time to Die

The great pioneering company and global business that H.R. MacMillan built, MacMillan Bloedel Limited, continued to evolve after his retirement. MacMillan's name is no longer identified with the business following its sale to Weyerhaeuser, an American-based lumber giant. Historically, most Canadian businesses remain under one title or in one family for only one or two generations. In the last decade of the 20th century, the same was true of the great company that MacMillan had founded.

While MacMillan's name vanished from the company, other highly visible aspects of the community will carry his name in perpetuity. Today, there are many deeply rooted community institutions that claim his name and continue to remember him with gratitude: **The MacMillan Planetarium**, now known as the **H.R. MacMillan Space Centre**, in Vancouver; **The H.R. MacMillan Gallery of Whales** and the **H.R. MacMillan Tropical Gallery** (both of which are in the Vancouver Aquarium); **H.R. MacMillan Park** in Chemainus; the **MacMillan Forestry Building** and its **MacMillan Library** at the University of British Columbia; **MacMillan Park** at Cathedral Grove on Vancouver Island; and the **H.R. MacMillan Endowment for Theological Education** and the **H.R. MacMillan Theological Library** at the Vancouver School of Theology. Hundreds of people, if not thousands, have benefited, and continue to benefit, from his generous gifts. Furthermore, future generations will appreciate MacMillan's systematic, focused, and intentional initiation of philanthropy in British Columbia. If the Vancouver Foundation is correct in its estimate that MacMillan gave over one hundred million dollars to worthy causes in British Columbia, then his gifts have far exceeded his own ten percent rule of life.[1]

Although British Columbians remember MacMillan for his generosity, he did not give financial gifts and endowments solely to be remembered for doing so. He supported a variety of institutions simply because they represented beliefs that he considered to be important. They were an extension of his core values and his ability to pass on financial resources to them gave him great pleasure. "It is more blessed to give, than to receive," was his mother's advice to him when he was ten years old and he never forgot it. "The way to enjoy yourself," she said, "is to do all you can to make other people enjoy themselves."[2] His endowments were a grateful celebration of the good fortune he had received in his own lifetime: family, faith, hope and love, education, opportunity, renewed health, Canada itself, and particularly, his beloved British Columbia. He gave in response to his own blessings but his gifts also demonstrated his faith in God and community

Bidding farewell

The funeral of Harvey Reginald MacMillan took place on February 12, 1976, at Christ Church Cathedral in Vancouver. The Cathedral was appropriately decorated with Douglas fir seedlings brought in by his family. One person who attended said, "The decorations in the church gave the impression that the whole congregation had assembled in a small grove deep in a dense old-growth forest!"[3]

Archbishop Godfrey Gower came out of retirement to speak to the congregation that filled the church. "Whatever was offered H.R. MacMillan took, and shaped a pattern of his own making," he said, "and he stamped it with the mark of his own individuality."[4] His philanthropy not only revealed his personal generosity, but he set a pattern of benefaction for individuals and corporations to emulate.

His extensive reading nourished his inner spirit and his friends and pastors complemented his passion for literature. He had a profound appreciation of the mystery of creation, though he would be the first to admit that he, like all people, did not have answers to the eternal quest of the meaning of life. For MacMillan, the greatest contribution anyone can make to the world is to become its servant, faithfully using the resources and talents with which one is gifted. This meant generously sharing some of his financial resources, not only with his family, but also for the benefit of the larger community. His vision and generosity were without precedent in the province. The legacy he left to his province, and the example he set, planted the seeds of philanthropy in British

Columbia and triggered an avalanche of like-minded successors who continue to endow the arts, sciences, humanities, and theological pursuits.

Twenty years before his death, MacMillan reflected openly on life and death. "God is very wise," he said, "when a man has lived a long while and has learned a great deal about how to do things, God arranges to take him away. It is a good rule."[5] MacMillan understood life to be a part of a divine plan, a cosmic order, a rule that shapes all human aspirations and actions.

H.R. MacMillan Park, Chemainus, BC

A collage of photos of buildings and institutions bearing the MacMillan name.

The same week MacMillan's funeral was held in Vancouver, Father Damasus Payne, in his sermon to the Poor Clare Sisters in Mission, BC spoke of him in positive and complimentary terms:

> He stood taller than the trees he dealt with. Some people won't picture him in such a glowing light. They called him a tiger. A hunter, a true hunter would be a better term. True hunters respect the game they stalk, because they are not alone in the woods. They see the signs of God's presence everywhere and know that he is running with them encouraging them to "Fight the good fight and take the prize."
>
> Having taken every prize within sight, I think that Harvey MacMillan had the wisdom to see that there is a better prize within the grasp of all. I think he was patiently and humbly waiting for the Lord to come and show him the world that He had conquered and wills to share with us.
>
> This is the crowning grace for a man who has run his course and won his prize – the grace to willingly and eagerly put everything aside so that he can receive something better.[6]

For three hundred years, fifty-five Wren churches dotted the skyline of London as testimony to a master artist of stone and symbol. Scattered on the social and institutional landscapes of British Columbia are dozens of institutions that remind us of MacMillan's challenge to us, "to build upon the heritage of the past with courage, realism, and unselfishness."[7]

Sadly, MacMillan's dream to place a Christopher Wren spire on the campus of the University of British Columbia or in the heart of Vancouver has yet to be realized. Nevertheless, in British Columbia, there could be no truer epitaph commemorating MacMillan than the inscription over the interior of the north door in St. Paul's Cathedral in London, which was written by Wren's son:

SI MONUMENTUM REQUIRIS, CIRCUMSPICE

'If you would see his monument, look around'

A legend in his own time

Today, Harvey Reginald MacMillan is remembered for something of far greater importance than his business and professional achievements, great though these were. At its core, his lasting legacy has a profound

Lisa Southam (left) and her mother Jean MacMillan Southam are seen with Allan Dewar (right), formerly of MacMillan Bloedel Limited, with sculptor Cheng She Ren (Arthur) at the unveiling of a statue of H.R. MacMillan in Chemainus in 1999.
Photo Credit Ying Cheng

spiritual dimension. In him, we find the mix of high ideals, values, and beliefs that led to him back to the community that gave so much to him. His is a legacy of example, a notable British Columbia precedent.

Long before philanthropy became widespread in British Columbia, MacMillan began to share his wealth. His philanthropic gifts far exceeded the gifts of his contemporaries and he set a pace rarely matched or exceeded. His rule of life drove him to give back to the community, whether in large or small gifts, in support of its religious, social, and educational institutions. In his lifetime, he was legendary for his benefactions. At the time of his death, the Governor General of Canada, Jules Léger, penned a letter of condolence to MacMillan's daughters, Marion and Jean. He wrote, "The contributions your father has made to the development of British Columbia are incalculable, of course, and his philanthropy has become a legend."[8]

A legend beyond death

MacMillan, in his thirteen-page will completed on July 27, 1973, has the final word about his commitment to philanthropy. His rule of life applied after his death – he specified that ten percent of the net value of his estate be set aside and invested in a separate trust fund. The principal and interest are to support "any programme designed to assist the Native people of northern Canada to advance their own welfare by participation in the development of the natural resources of northern Canada including the granting of scholarships and bursaries to the Native people for study and training at the University of British Columbia." He directed that scholarships and bursaries be available to those students whose parents are unable to provide the necessary financial support to complete their education, particularly those students who are likely to make a reasonable contribution to the future development of Canada in industry, agriculture, trade, science, or the professions. Conscious of the importance of developing natural resources lying north of the 60th parallel, he directed that studies be made in this area to benefit the Canadian economy but that due care be taken to protect its environment and Native inhabitants.

Another ten percent

In addition to the trust mentioned above, MacMillan directed that yet another ten percent of the net value of his estate be given to those charitable organizations that are exempt from any succession duties and whose works are exclusively within the province of British Columbia. The choice of the beneficiary in this case was to be left to the discretion of his trustees.

The recipients of MacMillan's posthumous gifts vary widely but are entirely in keeping with his lifetime interests. For example, he left his entire collection of rare books to UBC, and he left money to each of the fifteen small-town libraries on Vancouver Island. In addition, he directed that funding be disbursed to professional organizations, churches, and universities: the Empire Forestry Association; the Royal

Geographical Society; the Dean and Chapter of Westminster Abbey, London; St. Paul's Cathedral, London; the University of Oxford, and the University of Cambridge. Finally, on a more personal level, he directed that a lifetime monthly stipend be paid to his gardeners, housekeepers, and to his secretary.

Canada honours MacMillan

Both in his lifetime and after his death, MacMillan's name and gifts became legendary. Once again, Canada, his much beloved nation, honoured him. Although he had already been given the title Commander of the Order of Canada, and had later been made a Companion of the Order of Canada, the Government of Canada now presented MacMillan with a title rarely given to Canadians, and then only posthumously. In 1999, Canada named him: "A Person of National Historic Significance." At that time, there were approximately six hundred Canadians so named, including the founders of Canada, and of its provinces, and other historically significant people.

On March 29th, at the Forest Sciences Centre, on the east side of Main Mall of the University of British Columbia, across from the MacMillan Building, the Honourable Andrew Mitchell, Secretary of State (parks) and Member of Parliament for Vancouver-Quadra, unveiled a plaque honouring MacMillan. In announcing the event, the Heritage Canada press release characterized MacMillan as, "a man of generosity who donated to charities and special interest groups." The Honourable Mr. Mitchell read the citation of Canada, newly cast in bronze that reads in part: "Pioneer conservationist, industrialist, and philanthropist." Posthumously, a grateful country accorded Harvey Reginald MacMillan its highest title and designation.

Canada's acknowledgement of H.R. MacMillan as a person of national historic significance would have pleased him greatly. He had dedicated his life to the highest ideals valued by citizens of this country and, to an extraordinary degree, in the then-emerging field of Canadian philanthropy. The honour was a fitting tribute to British Columbia's father of philanthropy.

Harvey Reginald MacMillan
(1885-1976)

Pioneer conservationist, industrialist, and philanthropist, MacMillan was educated at Yale as one of Canada's first professional foresters. He became British Columbia's Chief Forester in 1912 and in 1919 formed his own company to market lumber abroad, participating in a tremendous expansion of the industry. When his company merged with Bloedel, Steward and Welch in 1951, it became one of Canada's largest forest enterprises. His efforts in organizing forest production in both world wars were recognized when he was made Commander of the Order of the British Empire.

Appendix I

APPENDIX 2

Davis, Hossie, Campbell, Brazier & McLorg
Barristers & Solicitors

TELEPHONE MUTUAL 5-8521
CABLE ADDRESS DAMARELL
CODES { A 1 TELEGRAPHIC / WESTERN UNION / BENTLEY'S / SLATER'S

D. N. HOSSIE, Q.C. (1962)
E. M. C. MCLORG
A. J. F. JOHNSON
D. B. D. HOSSIE
B. P. M. BYRNE
J. G. WATSON
J. R. MACKAY
D. M. JOHNSTON

A. T. R. CAMPBELL, Q.C.
A. W. FISHER
J. G. ALLEY
H. C. MILLHAM
G. K. FUJISAWA
A. W. SUTHERLAND
W. J. WRIGHT

C. W. BRAZIER, Q.C.
R. J. MCMASTER
A. B. FERRIS
E. S. THORNE
D. W. SHAW
D. H. PATERSON
M. R. TAYLOR
L. R. DESFOSSES

14th Floor, The Burrard Building
1030 West Georgia Street
Vancouver 5, B.C.

REPLY ATTENTION OF Mr. Brazier
FILE NO

January 28, 1965

The Right Rev. G.P. Gower, B.A., D.D.,
President,
Anglican Theological College of British Columbia,
Room 818,837 West Hastings St.,
Vancouver 1, B. C.

Dear Bishop Gower:

I am instructed by Mr. H.R. MacMillan to advise you that he has executed a Deed of Settlement establishing an endowment fund for the Anglican Theological College of approximately One Million Dollars in the hope and expectation that the Governing Body of the College will accept and use it as nearly as possible according to his wishes as expressed in the Deed of Settlement. The gift consists of 30,000 common shares of MacMillan, Bloedel and Powell River Limited and upon acceptance of the gift in accordance with the terms thereof by the Governing Body of the College, the shares will be held in trust by The Canada Trust Company for the College.

I am enclosing herewith the original and two copies of the Deed of Settlement signed by Mr. MacMillan and The Canada Trust Company and would ask you to execute the same under the corporate seal of the College and return the two copies to me at your earliest convenience, retaining the original for your records. Would you also have the appropriate officer of the College make arrangements with the Trust Company for the administration of the fund.

I do request of you that no publicity be given to the fact of the gift by the College or any official thereof until a press release has been made on behalf of Mr. MacMillan. It is expected that this will be done sometime during the week of February 1st.

Yours truly,

C. W. Brazier

CWB:MH

Appendix II

DEED OF SETTLEMENT *

THIS INDENTURE made the 27th day of January, 1965

BETWEEN: HARVEY REGINALD MacMILLAN, of 3741 Hudson Street, in the City of Vancouver, In the Province of British Columbia,

(hereinafter called the "Settlor")

OF THE FIRST PART

AND: THE CANADA TRUST COMPANY, a body corporate, of 901 West Pender Street, in the City and Province aforesaid,

AND: THE ANGLICAN THEOLOGICAL COLLEGE BRITISH COLUMBIA, of 6060 Chancellor Blvd., In the City of Vancouver

THE UNION COLLEGE OF BRITISH COLUMBIA, of 6000 Chancellor Blvd., In the City of Vancouver

(hereinafter called the "beneficiary")

WHEREAS the Settlor is desirous of settling on the Trustee the property hereinafter mentioned in trust for the Beneficiary to be used by the Beneficiary for the purposes hereinafter defined;

---

* Identical and separate Deeds of Settlement were drawn for, and signed individually by, each College. The names of the signatories of both Colleges are cited in this record.

NOW THEREFOR THIS INDENTURE WITNESSETH that"

1. The Settlor does hereby give, transfer and deliver absolutely to the Trustee Thirty Thousand (30,000) Ordinary shares of the capital stock of MacMillan, Bloedel and Powell River Limited (hereinafter referred to as "MB&PR") without nominal or par value with duly executed transfers thereof in favour of the Trustee upon trust to stand possessed of the said shares and all income hereafter earned therefrom to pay the said income and the capital from time to time realized upon any sale of the said shares or any investments substituted therefor to the Beneficiary to develop a programme of research, scholarship and training and in particular to afford members of the teaching staff of the Beneficiary opportunity for advanced study in larger centres, to establish fellowships in order to attract the services of visiting lecturers expert in their field, to offer post graduate courses, to afford opportunity for studies abroad for those showing competence in special fields, to offer short term refresher courses for the Clergy throughout the Province of British Columbia and to undertake such other projects of a similar nature as the Governing Body of the Beneficiary may decide upon from time to time, it being the wish of the Settlor that in the advanced studies and training undertaken hereunder there shall be the greatest possible degree of cooperation with other denominations, including exchange of personnel, results of research and of study.

2. The said income and capital of the trust property shall be applied to the purposes aforesaid as the Governing Body of the Beneficiary shall from time to time determine, it being the wish of the Settlor, however, that the capital of the said fund shall remain intact as a capital endowment fund and that the income only shall be employed for the purposes aforesaid but nothing in this deed shall fetter the discretion of the Governing Body of the Beneficiary to apply the whole or any part of the capital of the trust fund to the attainment of the purposes aforesaid.

3. The Trustee shall, subject to what follows, have power to sell or cause to be sold all or any of the shares of MB&PR constituting the trust fund hereby created and to invest and re-invest from time to time the proceeds of any such sale in such securities as are authorized by any statute of Canada for the investment of funds by life insurance companies.

   PROVIDED HOWEVER so long as the trust fund shall comprise shares of MB&PR thirty (30) days' notice in writing of the Trustee's intention to sell or alienate such shares. Within the said period of thirty (30) days' notice the said Chairman of the Board of MB&PR, or such officer or director of

MB&PR as he may appoint for the purpose, shall have the right to nominate a purchaser or purchasers or alienee or alienees of the said shares at the price and upon terms which the Trustee is willing to accept from any other purchaser or alliance.

4. The gift, transfer and delivery of the aforesaid shares of MB&PR to the Trustee and this deed are irrevocable by the Settlor.

5. The Trustee hereby acknowledges receipt of the aforesaid thirty Thousand (30,000) shares of MB&PR and agrees to stand possessed thereof upon the trusts hereby declared.

6. The Beneficiary hereby agrees to accept the benefit hereof and to use its best endeavors to apply the income of the trust fund and any part of the capital of the trust fund received by it for the purposes set forth and defined in paragraph 1 hereof.

IN WITNESS WHEREOF the Settlor has hereunto set his hand and seal and the Trustee and the Beneficiary have hereunto caused their respective corporate seals to be affixed in the presence of their respective duly authorized officers in that behalf all as of the day and in the year first above written.

SIGNED, SEALED AND DELIVERED	)	
by HARVEY REGINALD MacMILLAN,	)	***HRMacMILLAN***
the Settlor, in the presence of:	)	
	)	
A.M. Brazier	)	
The Corporate Seal of THE	)	The Canada Trust Company

Notes

Abbreviations:

1. **UBC/SC** refers to University of British Columbia and the MacMillan Personal Papers held in the Special Collections of the Library. The first number, records the box number. After the period, the number refers to the file within the box.
2. **VST** refers to the Archives at the Vancouver School of Theology using the same Number order/system as above.
3. **ATC** refers to the Archives of the Anglican Theological College held in the Vancouver School of Theology. The same number system as that of VST applies.
4. **UC** refers to Union College Archives held in the Vancouver School of Theology with same number system.

The End is the Beginning ...

[1] *Globe and Mail,* February 16, 1976.
[2] The Vancouver Sun, February 10, 1976.
[3] UBC/SC 82.7.
[4] Ken Drushka, *HR, A Biography of H.R. MacMillan,* Harbour Publishing, 1995, p.12
[5] Roland Wild, *Maclean's* magazine, March 15, 1948, p. 13.
[6] Walt Whitman, *Song of Myself,* 1900

Chapter One A Time To Be Born

[1] Arthur G. Dorland, *The Quakers in Canada – A History,* Ryerson Press, Toronto, 1968, p.15.
[2] *Ibid.,* pp. 104-11 gives an account of David Willson's activities in the Society of Friends, including the building of the Sharon Temple.
[3] *Ibid.,* p.104.

[4] The incident is described briefly in a Dec. 4,1961 letter from MacMillan to Dr. W.S. Taylor, UBC/SC 4.10.

[5] *Sing One: A Hymn Supplement,* Anglican Church of Canada, 1980.

[6] Arthur G. Dorland, *Op. cit.,* p.110.

[7] *Ibid.,* p. 111.

[8] Donald MacKay, *Empire of Wood: The MacMillan Bloedel Story,* University of Washington Press, Seattle, 1982, p. 28.

[9] Ken Drushka, *Op. cit.,* p. 22: UBC/SC 84-5 MacMillan letters to Mrs. D.J. Sinclair, August 25,1972 and May 31,1972.

[10] UBC/SC 83.1.

[11] Rufus M. Jones, *Later Periods of Quakerism,* Vol. 1, p.36.

[12] UBC/SC 34.12 Quoted in Drushka, *Op. cit.,* p. 24.

[13] Robert MacRae interview with Dr. William C. Gibson, a friend of H.R. MacMillan, May 4, 2003.

Chapter Two A Time to Prepare

[1] A. Harris Mitchell, "MacMillan Story," *Canadian Business,* November 1949, p. 98

[2] Ken Drushka, *HR., Op.cit.,* p. 32.

[3] A. Harris Mitchell, *Canadian Business,* "The MacMillan Story," November 1949.

[4] Ken Drushka, *Op. cit.,* p. 31.

[5] *Ibid.,* p. 36.

[6] *Ibid.,* p. 32.

[7] *Ibid.,* H.R.M. quoted, p. 43.

[8] *Ibid.,* p. 44.

[9] *Ibid.,* p. 44.

[10] UBC/SC 24.1.

[11] *Maclean's* magazine, March 15, 1948, p.46.

[12] A. Harris Mitchell, *Op. cit.,* p. 23.

Chapter Three A Time to Heal

[1] Harvey Southam, an unpublished biography.

[2] *The Canadian Magazine,* "Fighting the White Plague," p. 74, quoted in Ken Drushka, *Op. cit.,* p. 57.

[3] Ken Drushka, *HR, Op. cit.,* p.57.

[4] MacMillan letter to Mazo de la Roche, November 11, 1937, quoted in Drushka, *Op.cit.*

[5] MacMillan to Edna Mulloy, December 24, 1910, quilted in Drushka, *Op.cit.*
[6] Donald MacKay, *Op. cit.*, p.33.
[7] *Maclean's* magazine *Op. cit.*, p. 48.
[8] *The Aurora Banner*, August 4, 1911, p. 4.
[9] Michael Bliss, "Northern Wealth," *Canadian Courier*, December 1994/January 1995, p. 4.

Chapter Four A Time to Mature

[1] Donald MacKay, *Op. cit.*, p. 28.
[2] According to Plutarch from a letter written by Caesar announcing his victory at Zela concluding the Pontis campaign, May 47 B.C.
[3] Ken Drushka interview with the Hon. H.P. Bell-Irving, March 1995, See Druska, *Ibid.*, p.130.
[4] Luke 1:37 and Matthew 17:20.
[5] Donald MacKay, *Op. cit.*, p. 35.
[6] *Daily Colonist,* Victoria, July 18,1912, p.1.
[7] Ken Drushka, *Op. cit.*, p. 73.
[8] Harvey Southam, *Op. cit.*
[9] Ken Drushka, *Op. cit.*, p. 79.
[10] *Ibid.*, p. 79.
[11] *Ibid.*, p. 80.
[12] *Ibid.*, p. 80.
[13] Donald MacKay, *Op. cit.*, p.37.
[14] Robert Slater, *Telegraph Code to ensure SECRECY in the Transmission of Telegrams,* Simpkin, Marshall, Hamilton, Kent & Co. London 1916. The location in MacMillan's office was identified by Dorothy Dee, Secretary, when she sorted his papers for transfer to the UBC's Special Collections.
[15] UBC/SC 39.19.
[16] UBC/SC 36.19.
[17] Donald MacKay, *Op. cit.*, p. 39.
[18] Ken Drushka, *Op. cit.*, p. 100.
[19] W.H. Olsen, *Water over the Wheel,* Chemainus Crofton and District Chamber of Commerce, 1963, p.161.
[20] Donald MacKay, *Op.cit.*, p. 40.
[21] Frank Walden, "The H.R. Story", *The Vancouver Sun,* October 30, 1964, Business Section.
[22] Donald MacKay, *Op. cit.*, p. 41.
[23] Ken Drushka, *Op. cit.*, p. 111.

Chapter Five A Time to Invest

[1] Donald MacKay, *Op. cit.*, p. 43.
[2] Robert MacRae interview with Mary-June Pettyfer, secretary to Montague Meyer, June 27,1998.
[3] Ken Drushka, *Op.cit.*, p.43.
[4] *Ibid.*, p.43.
[5] *Maclean's, Op. cit.*, p.48.
[6] *Ibid., Op. cit.*, p.49.
[7] MacMillan, H.R., *Looking Back On This Business*, December 8, 1958, UBC/SC 82.1.
[8] *Ibid.*, p. 8.
[9] Murray Newman, Murray, *Life in a Fishbowl*, Douglas and McIntyre, Vancouver/Toronto, 1994, p. 38.
[10] *Ibid.*, p. 41.
[11] UBC/SC Memorabilia Box 5.
[12] *Ibid.*
[13] UBC/SC Memorabilia, Scrapbook, Box 2.
[14] *Ibid.*
[15] *Saturday Night*, February 22, 1941, p.11.
[16] UBC/SC 90.13. p. 1.
[17] *Ibid.*, p. 3.
[18] *Ibid.*, p. 8.
[19] *Time*, April 14, 1941.
[20] MacMillan, H.R., *Looking Back On This Business*, 82.5 p.2.
[21] *Ibid.*, p.3

Chapter Six A Time to Share

[1] Chalmers United Church, Archives, Vancouver School of Theology.
[2] Ken Drushka, *Op .cit.*, p. 144.
[3] *Ibid.*, p. 349.
[4] *Ibid.*, p. 255.
[5] *Ibid.*, p. 351.
[6] *Ibid.*, p. 351.
[7] *Ibid.*, p.12.
[8] UBC/SC 65.13.
[9] Ken Drushka, *Op. cit.*, p. 353.
[10] UBC/SC 65.7.
[11] *Ibid.*

[12] *The Vancouver Sun*, August 8, 1962.
[13] Ken Drushka, *Op. cit.*, p. 352.
[14] *UBC Gifts, Grants and Bequests*, LE 3, B794, B7: 1959-66
[15] *Creative Giving: a listing of Private Gifts, Grants and Bequests to UBC*, April 1971

Chapter Seven A Time to Retire

[1] H.R. MacMillan, *Looking Back On This Business, Op. cit.*
[2] UBC/SC Memorabilia, Box 5.
[3] *Ibid.*
[4] UBC/SC 82.7, Letter of Condolence to the daughters of MacMillan.
[5] UBC/SC 40-42: H.R. MacMillan's Personal Papers, records of grants.
[6] *Ibid.*
[7] Robert MacRae interview with W.S. Taylor, December 3,1997. Tape recording in VST Archives.
[8] William S. Taylor, *Step by Step by Step: An Anecdotal History of the Growth of Union College, 1948-1971*, VST, 1993, p.38.
[9] *The Story of Union College*, The Board of Governors, Union College, Vancouver, 1972, p.17.
[10] *Ibid.*, p. 39.
[11] VST/Archives, Principal's Papers, letter of F.K. Brown, August 31,1978.
[12] Robert MacRae interview of W.S. Taylor, *Op. Cit.*
[13] UBC/SC 46.14.
[14] H.R. MacMillan, to W.S. Taylor, January 12,1959, VST 87.6, File RG 03.
[15] UBC/SC 65.8.
[16] VST Principal's Papers, December 27,1960.
[17] UBC/SC 41.10.
[18] *Ibid.*
[19] *Ibid.*
[20] UBC/SC 65.8.
[21] *Ibid.*
[22] *The Acts of the Apostles 10:47.*
[23] This award was repeated in January 1966 to cover a further three years of the program.
[24] UBC/SC 66.23.
[25] *Ibid.*
[26] *Ibid.*
[27] UBC/SC 65.14.
[28] UBC/SC 65.8.
[29] *Ibid.*

Chapter 8 A Time to Endow

[1] See Appendix I.
[2] This memorandum cannot be found.
[3] VST 1.4, Anglican Theological College, Minutes of Board of Governors, February 3, 1965.
[4] *Ibid.*
[5] VST Union College Archives, Board of Directors, Minutes, January 30.1965.
[6] *The Province,* February 10, 1965.
[7] VST, 1.4, Anglican Theological College, Board of Governors, Records.
[8] UBC/SC 46.14.
[9] John B. Macdonald, Annual President's Report to UBC, UBC Archives.
[10] Robert MacRae interview with W.S. Taylor, *Op. cit.*
[11] *Ibid.*
[12] *An Act to Incorporate the Vancouver School of Theology,* Government of British Columbia, April 2, 1971.

Chapter 9 A Time to Dream

[1] Donald MacKay, *Op. cit.*, p. 43.
[2] *Ibid.*, p. 43.
[3] UBC/SC 82.5 Letters of condolence.
[4] Maclean's, Op. cit., p. 46.
[5] February 20, 1976.
[6] Christopher G. Curtis, *The Canadian Encyclopedia,* Hurtig, Edmonton, 1985, Vol. II, p. 1067.
[7] Ken Drushka, *Op.cit.*, p. 274.
[8] Reginald H. Roy, *The Vancouver Club: First Century 1889-1989,* Friesen and Sons, Cloverdale, BC, p. 266.
[9] UBC/SC Box 5, Miscellaneous Envelopes: Notes on "Ion's Spire" prepared by Rev. P.B. Clayton, Wakefield House, 41 Trinity Square, London, EC.3.
[10] UBC/SC 43.4, Gibson to MacMillan, August 26, 1967.
[11] William C. Gibson, Medical Comets: Scholarly Contributions by Medical Undergraduates, UBC Alumni Association, 1997 p. 65.
[12] Robert MacRae interview with W.C. Gibson, May 12, 2003.
[13] UBC/SC 43.4.
[14] UBC/SC 41.11. *Proposal to Vancouver Foundation.*
[15] *Ibid.* Gibson to MacMillan, November 27, 1972.
[16] *Ibid.* postcard.

[17] *Canadian Profiles*, Vol. One, George Lonn, Pitt Publishing Co., p. 149-50, UBC/SC, 81.7.

[18] UBC/SC 43.12.

Chapter Ten A Time to Die

[1] Vancouver Foundation brochure, *Welcome to the Vancouver Foundation*, March 2003.

[2] Ken Drushka, *Op cit.*, p. 25.

[3] T. Herbert O'Driscoll, Dean of Christ Church Cathedral, Vancouver 1968-82, personal interview with Robert MacRae.

[4] Donald McKay, *Op. cit.*, p. 308.

[5] *People in Management, BC Forest Industries Year Book*, 1965, p.17.

[6] UBC/SC 82.7 Letters of condolence to the daughters of MacMillan.

[7] UBC/SC 43.12.

[8] UBC/SC 82.7.

Credits

Photographs, illustrations and signatures (s)

Archives and Special Collections, University of British Columbia: pp. 14, 16, 19, 22, 24, 26 (s), 30, 32, 33, 36, 49, 51, 61, 64, 65, 66, 67, 67 (s), 68, 69, 70, 70 (s), 71 (s), 72, 73, 74 (s), 76, 83, 89, 94, 96, 96 (s), 98, 99, 101, 101 (s), 103, 103 (s), 111, 116, 116 (s), 117, 130, 132.

University of Guelph: pp. 26, 28, 30.

Rare Book Library, Vancouver School of Theology: pp. 105, 109.

Anglican Diocese of New Westminster: pp. 114, 114 (s).

All Hallows by the Tower, London: p. 125.

Archives, Christ Church Cathedral, Vancouver: p. 134.

Robert MacRae: pp. 18, 59, 113, 115, 126, 127, 129, 139, 144.

Archives, Vancouver School of Theology: pp. 82, 104.

Vancouver Foundation: p. 86.

Archives, City of Vancouver: p. 64 (s).

Ying Cheng, Richmond, BC: p. 141.

Front cover photo: Phoebe MacRae – the statue of H.R. MacMillan overlooking the dock and original site of the lumberyard of the Victoria Lumber Company, Chemainus, BC.